SCIENTIFIC METHOD

Applications in Failure Investigation and Forensic Science

INTERNATIONAL FORENSIC SCIENCE AND INVESTIGATION SERIES
Series Editor: Max Houck

Firearms, the Law and Forensic Ballistics
T A Warlow
ISBN 9780748404322
1996

Scientific Examination of Documents: methods and techniques, 2nd edition
D Ellen
ISBN 9780748405800
1997

Forensic Investigation of Explosions
A Beveridge
ISBN 9780748405657
1998

Forensic Examination of Human Hair
J Robertson
ISBN 9780748405671
1999

Forensic Examination of Fibres, 2nd edition
J Robertson and M Grieve
ISBN 9780748408160
1999

Forensic Examination of Glass and Paint: analysis and interpretation
B Caddy
ISBN 9780748405794
2001

Forensic Speaker Identification
P Rose
ISBN 9780415271827
2002

Bitemark Evidence
B J Dorion
ISBN 9780824754143
2004

The Practice of Crime Scene Investigation
J Horswell
ISBN 9780748406098
2004

Fire Investigation
N Nic Daéid
ISBN 9780415248914
2004

Fingerprints and Other Ridge Skin Impressions
C Champod, C J Lennard, P Margot, and M Stoilovic
ISBN 9780415271752
2004

Firearms, the Law, and Forensic Ballistics, Second Edition
Tom Warlow
ISBN 9780415316019
2004

Forensic Computer Crime Investigation
Thomas A Johnson
ISBN 9780824724351
2005

Analytical and Practical Aspects of Drug Testing in Hair
Pascal Kintz
ISBN 9780849364501
2006

Nonhuman DNA Typing: theory and casework applications
Heather M Coyle
ISBN 9780824725938
2007

Chemical Analysis of Firearms, Ammunition, and Gunshot Residue
James Smyth Wallace
ISBN 9781420069662
2008

Forensic Science in Wildlife Investigations
Adrian Linacre
ISBN 9780849304101
2009

Scientific Method: applications in failure investigation and forensic science
Randall K. Noon
ISBN 9781420092806
2009

SCIENTIFIC METHOD

Applications in Failure Investigation and Forensic Science

Randall K. Noon

CRC Press is an imprint of the
Taylor & Francis Group, an **informa** business

Cover art courtesy of NASA.

CRC Press
Taylor & Francis Group
6000 Broken Sound Parkway NW, Suite 300
Boca Raton, FL 33487-2742

Printed in the United States of America on acid-free paper
10 9 8 7 6 5 4 3 2 1

International Standard Book Number-13: 978-1-4200-9280-6 (Hardcover)

Library of Congress Cataloging-in-Publication Data

Noon, Randall.
Scientific method : applications in failure investigation and forensic science / Randall K. Noon.
p. cm. -- (International forensic science and investigation series)
Includes bibliographical references and index.
ISBN 978-1-4200-9280-6 (hardcover : alk. paper)
1. Science--Methodology. 2. Reasoning. 3. Accident investigation. 4. Criminal investigation. 5. Forensic engineering. I. Title. II. Series.

Q175.32.R45N66 2009
620--dc22 2009006365

Visit the Taylor & Francis Web site at
http://www.taylorandfrancis.com

and the CRC Press Web site at
http://www.crcpress.com

The Fog Went Thud

Carl thinks cats come in like a quiet fog,
And sit on silent haunches surveying the realm.
These are certainly not my cats,
Who drop from forbidden dining room tables
And thump like bowling balls,
Who claw upholstery chasing invisible prey,
Or skittle about battling paper mice.
No fog I know
Percolates like a straight-eight Ford,
Knows how to cozy on a chilly evening,
And stares at fairies lurking in empty corners.
Carl knew Lincoln better than anyone,
And could turn a phrase with the best,
But I don't think he ever really had a cat.

For Leslie — who misses Elwood

Table of Contents

International Forensic Science Series

The modern forensic world is shrinking. Forensic colleagues are no longer just within a laboratory but across the world. E-mails come in from London, Ohio and London, England. Forensic journal articles are read in Peoria, Illinois and Pretoria, South Africa. Mass disasters bring forensic experts together from all over the world.

The modern forensic world is expanding. Forensic scientists travel around the world to attend international meetings. Students graduate from forensic science educational programs in record numbers. Forensic literature—articles, books, and reports—grows in size, complexity, and depth.

Forensic science is a unique mix of science, law, and management. It faces challenges like no other discipline. Legal decisions and new laws force forensic science to adapt methods, change protocols, and develop new sciences. The rigors of research and the vagaries of the nature of evidence create vexing problems with complex answers. Greater demand for forensic services pressures managers to do more with resources that are either inadequate or overwhelming. Forensic science is an exciting, multidisciplinary profession with a nearly unlimited set of challenges to be embraced. The profession is also global in scope—whether a forensic scientist works in Chicago or Shanghai, the same challenges are often encountered.

The International Forensic Science Series is intended to embrace those challenges through innovative books that provide reference, learning, and methods. If forensic science is to stand next to biology, chemistry, physics, geology, and the other natural sciences, its practitioners must be able to articulate the fundamental principles and theories of forensic science and not simply follow procedural steps in manuals. Each book broadens forensic knowledge while deepening our understanding of the application of that knowledge. It is an honor to be the editor of the Taylor & Francis International Forensic Science Series of books. I hope you find the series useful and informative.

Max M. Houck
Series Editor

Foreword

I have noted elsewhere that forensic science is a historical science,[1] and the same has been said by others.[2] That fact puts the practice of the profession at a certain disadvantage: A forensic scientist is always removed from events he is trying to reconstruct, and those events are themselves abstracted from the analysis the scientist conducts. Time moves in only one direction, sweeping with it many of the details one might otherwise need for a complete reconstruction.[3] The details that are left behind, what Locard called "traces"[4] and what I call "proxy data,"[5] are necessarily incomplete, occasionally vague, and differentially valued.[6] Nonetheless, these remnants are evidence in criminal cases. Forensic science attempts to connect these "dots" to reveal any relationships among people, places, and things that were involved in the criminal activity. This notionally suggests a reverse causality, that is, a piecing together of the remnants of the past activity to create a logical narrative that relates the events *in their purported original order* of more or less cause and effect. Again, the forensic scientist's abstractedness brackets what can and cannot be said about those past events.

The fact that one event regularly comes after another is often evidence that the prior caused the latter; throwing a stone at a window with sufficient force will cause the window to break. Other times, this is not the case—saying "Open Sesame!" before stepping through the grocery store's automatic door is not what causes it to open. As Fowler succinctly puts it:

> Thus while we would agree that the throwing of the stone (the prior event) causes the latter event (the breaking of the window) in this particular situation, we would not say that, for instance, day *caused* night, although we do perceive a regular sequence of night following day. In other words, we do in real life make a distinction between something being *post hoc* (merely following a previous event) and *propter hoc* (being *caused* by the previous event).[7]

We thus infer that a necessary connection exists between the stone and the window breaking and use the word *caused* to demonstrate that connection ("Throwing the stone caused the window to break"). The point that is often missed, however, is that the events that are chosen as "cause" and "effect" are really separate entities; there is no logical sufficiency to connect the two other than through our interpretation of the events. The law, or perhaps The Law, seeks direct one-to-one cause-and-effect results to clarify the trier of fact's decision. That is a legal necessity and not a scientific one. Hume indirectly made this distinction by stating that, for it to be successful, the scientific method must be based not on single cause–effect scenarios but on "collections of frequently observed event sequences" (Fowler[7], p. 62). Summary and analytical numerical indices deal with the collections Fowler notes. Given some population that is too large or too unwieldy to measure directly, samples and subsamples are collected and the aggregate data (Fowler's collections) are analyzed as indicators of the larger populations. Statistics and probability, therefore, are the basis for solid scientific interpretations.

Forensic science is only now elucidating its statistical basis for interpretations, with DNA being the first example. Interestingly, DNA may be the anomaly—and not the gold standard—of forensic science. Allele frequencies are really just class characteristics (more than one person can have the same allelic variant at a specific locus); their power comes from their independence, which allows the allele frequencies to be multiplied. Of course, *the forensic loci were chosen because they are independent.*[8] The rest of the forensic disciplines have not had this luxury and may not be afforded it; for example, how does one determine independence for features of a shoe tread intentionally designed for some end purpose like better traction for jogging? These areas of forensic analysis are only now wrestling with many fundamental and upsetting realizations, chief among them that the concept of individualization[9] may be philosophically and existentially valid but scientifically bankrupt.[10] Forensic science is now in the unenviable situation of playing catch-up with decades of practice to repair and renovate its philosophical foundations.

Noon's book is an excellent step in this process. Although it is oriented to forensic engineering, many of the ideas and principles apply to most if not all of forensic science. Well-written, concise, and illustrative, Noon has provided those of us with a notion of what the theory and philosophy of forensic science looks like with a wonderful stone with which to shore up the foundation of our discipline.

Max M. Houck
West Virginia University

References

1. Houck, M. M., and J. A. Siegel. 2006. *Fundamentals of forensic science.* San Diego: Academic Press.
2. Cleland, C. E. 2001. Historical science, experimental science, and the scientific method. *Geology* 29:987–90.
3. Davies, P. 2002. That mysterious flow. *Scientific American* 287:40–47.
4. Locard, E. 1939. *Manual of police techniques*, trans. Kathleen Brahney. 3rd ed. Paris: Payot.

5. Houck, M. M. 2006. Playing fast and loose with time and space: Statistics and forensic science. Paper presented at the Joint Statistical Meetings, Seattle.
6. Houck, M. M. 2006. Foreword—The grammar of forensic science. *Forensic Science Review* 18(2).
7. Fowler, W. S. 1962. *The development of the scientific method.* London: Pergammon Press, p. 62.
8. Jobling, M. A., and P. Gill. 2004. Encoded evidence: DNA in forensic analysis. *Nature* 5:739–52.
9. Kirk, P. 1963. Criminalistics. *Science* 140:367–70.
10. Saks, M. J., and J. J. Koehler. 2008. The individualization fallacy in forensic science evidence. *Vanderbilt Law Review* 61:199–219.

Preface

The photograph of the Space Shuttle *Challenger* on the cover, which was taken 13 seconds after lift-off, is a blunt reminder about why it is necessary to investigate failures, accidents, and crimes. Seven remarkable people died when the *Challenger* exploded just after launch. Seven families were grievously affected. Friends, neighbors, workmates, communities, and a nation mourned. National treasure was wasted. The time invested by thousands of people was wasted. The reputation of a nation, an agency, and a program was severely damaged. It was a very hard loss to bear.

To avoid such tragic losses of life, property, reputation, and treasure, we search for the reasons why these events occur. Not only is it expedient to do so, for many reasons, but it also satisfies a psychological requirement for closure. For those who are directly affected by such losses, it is difficult not to ask: Why did this happen?

Thus, we take the information we learn from an investigation and use it to devise ways, hopefully, to avert similar tragedies. If we ruthlessly apply logic and proven scientific principles in our investigation, we may be successful. If we allow biases to affect our logic and make mistakes understanding the evidence, we will likely fail.

The Space Shuttle *Challenger* is an example of this. The postaccident investigation and the follow-up implementation of corrective actions were done with great effort, expense, and determination by many dedicated people. However, the effort was flawed.

On February 1, 2003, another shuttle tragedy replayed itself across the television screens of the world. After a successful mission in space, the Space Shuttle *Columbia* reentered the atmosphere. During reentry, a hole burned through the port-side wing, where a heat-resistant tile had been damaged during lift-off. The space ship broke apart and another seven of our brightest and best died.

This book contains six subject chapters, a seventh case study chapter contributed by Jon J. Nordby, PhD, D-ABMDI, and an eighth chapter that is a short reading list for those who wish to read more about this book's subject. I have attempted to provide theory, historical context, and practical experience. I have deliberately not used a formal academic format. In writing the manuscript, I imagined that I was talking about the various topics to a good friend over a cup of coffee. In some cases, that is exactly how the topic first came to mind.

The best way to become proficient at almost anything is to do it often: practice, practice, practice. This is also true with failure, accident, and crime investigations. Studying the available literature, however, does reduce the amount of time necessary to become proficient. It also allows a person the opportunity to avoid the mistakes already made by others—and perhaps pioneer new ones.

To those considering working in this field, I recommend that the short reading list in the last chapter be considered. Take advantage of the free Internet downloads I have listed. If there is a good library nearby, request and read at least some of the books, monographs, and articles. I also suggest that a person work at least for a time with an experienced investigator in a mentor–pupil relationship. I subscribe to the "work in the store before you try to run one" philosophy. While there may exist better methods of learning, I don't know what they are. Besides, there are many little tricks of the trade, especially in dealing with administrative matters and people that are peculiar to each region and organization, which cannot easily be incorporated into a book intended for general readership.

Applying a method is not a substitute for basic subject matter knowledge, evidence, or facts. If there were a heart transplant failure case that needed to be investigated, I don't think a group of bricklayers would do as good a job in figuring out the issues as would a group of doctors and surgeons. Conversely, in a controversy involving masonry workmanship, I believe that a group of bricklayers would do a better job than a team of doctors and surgeons. Methodology is simply a framework within which subject matter knowledge, evidence, and facts are organized and assessed.

With respect to the *Challenger* and *Columbia* Space Shuttle tragedies, I invite the reader to check the last chapter, "Reading List," look up the *Columbia* Accident Investigation Board report, and read at least the executive summary contained in Volume 1. It is a free download from the Internet. I also invite the reader to review Chapter 5 in this book and decide which of the four pitfalls applied in that instance.

Lastly, I am indebted to Dr. Jon Nordby for providing an excellent case study. It not only ably demonstrates principles discussed in the rest of the book, but also is a good true-life detective story in its own right.

With those thoughts in mind, I hope the reader enjoys and benefits from the materials presented. I trust that one of you will improve upon my humble effort. I look forward to reading it, or perhaps you can tell me about it over a cup of coffee sometime.

RN

About the Author

Randall Noon is a licensed professional engineer who has been involved in failure analysis, root cause analysis, and forensic engineering for many years. Currently, Mr. Noon is a root cause team leader at a nuclear power plant. Previously, Mr. Noon contributed two chapters to the text *Forensic Science: An Introduction to Scientific and Investigative Techniques*, edited by Stuart James and Jon Nordby.

1 Introduction

One cool judgment is worth a thousand hasty counsels. The thing to be supplied is light, not heat.

Woodrow Wilson, January 29, 1916

General

This book describes and explains some of the investigative methods employed to determine why and how a particular event occurred. The usual terms of art to describe this activity include *root cause investigation*, *root cause analysis*, *failure analysis*, *forensic investigation*, *forensic engineering*, *accident reconstruction*, *failure reconstruction*, *failure investigation*, and so on. While there are different shades of meaning among the terms, and various trades and professional groups have their favorite terms, they all embrace determining how and why an event occurred.

Consequently, this book is fundamentally about investigating cause-and-effect relationships. Given the end result, that is, the effect, this book describes some of the systematic methods employed to determine the event's cause.

Usually the effect to be investigated is something bad: an accident, a failure, a catastrophic event, or a crime. Investigations, however, don't have to include only events that are bad. There is certainly value in understanding clearly how things are accomplished successfully. Unfortunately, while it is common to investigate the reasons why something bad has occurred, to determine who or what was at fault, the reasons for successful results are often taken for granted.

When a part, component, material, system, procedure, or management method appears to work well in one situation, but inexplicably fails when applied elsewhere, the reason it was successful in the first place was likely not clearly understood. The cause for success in the first case may have been presumed rather than proven by rigorous assessment of the evidence. Thus, the lack of accurate knowledge about why the part, component, material, system, procedure, or management method was successful in the first situation prevented potential failures in other applications from being predictable and avoidable.

Most failure or accident investigations begin at the end of the story: after the explosion, after the fire has been extinguished, after the collapse, and so on. It is at this point that people ask: How did this happen, and who or what is to blame for this? From this chronological endpoint, the investigator gathers, verifies, and assesses evidence to determine how and why the event came to occur.

In many instances, information about both the last event and the starting event is known reasonably well. Information about what occurred between these endpoints, however, is often unclear, confusing, and perhaps even contradictory. The intermediate points usually reveal themselves in a piecemeal fashion as evidence is gathered and assessed.

- Some evidence items, of course, are not time and date stamped for the convenience of the investigator. Their place in the event sequence must be deduced or inferred by statements, context, and logic.
- Then again, some items are indeed time and date stamped, such as bills of lading, credit card receipts, money machine receipts, e-mail, or postmarked letters. These items often serve as verifiable anchor points in a timeline.
- Some presumed evidence items are simply red herrings. In the end, they are found to have no significant relationship to the event being investigated. These items may at first appear to be related to the event, or someone may claim they are related to the event. Red herrings can cost an investigator valuable time and resources to validate that they are unimportant. However, if they are not properly investigated and found out, they can cause the final conclusions of the investigation to be significantly flawed.
- Some important evidence items may never be discovered at all. They may have been inadvertently discarded during the investigation or cleanup, lost, deliberately destroyed to conceal the true cause of the event by the perpetrator or accomplice, or perhaps destroyed by the event itself. Consider the paper match that ignites a forest fire.
- Some evidence may be misinterpreted by the investigator. Perhaps its significance is overlooked, overvalued, or not understood properly at all. For example, the significance of material fatigue may not be properly evaluated if the investigators cannot recognize what a fatigue fracture looks like.
- Some evidence items may be deliberate falsifications. Such items may not be a natural precursor, result, or consequence of the event. They may have been deliberately positioned at the accident scene, or within the evidence trail, to be discovered by the investigator. They are intended to mislead the investigator and cause him to reach an incorrect conclusion.

In any case, it is the investigator's job to assess, sort, and piece the evidentiary pieces into a logical, coherent story of how and why the event came to pass. Like solving a puzzle that has some missing pieces, and perhaps a few extra pieces thrown in from other puzzles, the investigator considers each piece of evidence many times and in different combinations to determine where it fits in the overall picture.

For example, in a car accident between cars A and B, the positions where the vehicles and the accident victims came to rest will usually be well known. This is the endpoint of the timeline. The relative positions of the vehicles and participants prior to the accident, that is, the starting point, may also be known. For example, car A was heading west on Elm Street with Joe and Jim in it, and car B was heading north on 2nd Street with Mr. and Mrs. Smith in it. The investigator then fills in what occurred between the start point and the endpoint on the timeline.

Witnesses may document some parts of the story in statements, and the participants may document some parts of the story in statements. Unfortunately, the various statements will likely conflict with one another, lack detail, contain convenient exaggerations or understatements, contain ill-formed conclusions or personal theories disguised as observations, or be incomplete. The gaps in information, the resolution of conflicting information, the degree of exaggeration and understatement, and the discrimination between ill-formed conclusions and actual observations are arbitrated by the physical evidence.

Often, some important condition completely unknown to any of the witnesses or participants may come to light if the investigation is thorough and unbiased. For example, a mechanical problem in one of the vehicles that was unknown to any witness or participant might have played a significant role in the accident, perhaps by slightly lengthening the stopping distance of the vehicle. Perhaps the timing of the traffic lights was inappropriate because not enough time for the yellow light was allowed. Perhaps the intersection itself was poorly designed for the amount of traffic being borne, or the speed limit was posted 10 mph too high. It is surprising how many "obvious" cause determinations based upon witness statements are readily overturned by just a modest amount of investigation into the physical evidence and facts.

Eyewitnesses usually do not see or perceive latent deficiencies. They are poor judges of time, distance, loudness, brightness, and temperature. They often focus on the immediate accident event, which is noisy and attention getting, to the exclusion of other events and circumstances that existed in the immediate area at the same time. These other events may have occurred with less fanfare but may have been more significant with respect to the causation of the accident event.

Houdini and Blackstone are famous for expertly creating diversions to pleasantly misdirect thousands of eyewitnesses all at the same time. Some audience members were so convinced by the stage illusions that they believed they had seen "the real thing." As these famous magicians knew:

- Eyewitnesses tend to see what they already understand and wish to see.
- They don't see what they don't understand and don't want to know.
- They tend to focus on what is bright, colorful, sexy, and noisy and ignore what is quiet, bland, and unobtrusive.

Eyewitnesses often do not directly report their observations. They tend to report their personal conclusions as they internally try to make sense of what they have observed. It is a natural tendency of people to try to make sense of disconnected events they have observed and "connect the dots." Over time, as they think about things and listen to the comments of friends and neighbors, they often report having seen things that they did not actually witness. Likewise, over time they may forget things that they may have actually witnessed. They remember what they think is interesting or what other people think is interesting, and forget what they think is unimportant.

And, of course, some witnesses simply lie. They lie perhaps for sport, perhaps for attention or self-amusement, or perhaps for the power over the situation that the lie can bring about. Often, however, they simply lie for personal gain.

What a Failure Investigator Does

Like a good journalist, a failure investigator endeavors to determine who, what, when, where, why, and how. When a particular failure, crime, or event has been explained, when the sequence of events has been determined, and when the main questions about causation have been answered, it is said that the event, failure, or crime has been solved or reconstructed.

To establish a sound basis for analysis, an investigator relies upon the actual physical evidence found at the scene and verifiable facts and records related to the matter. The investigator then applies accepted methodologies and well-proven scientific principles to organize and interpret the physical evidence and facts. The investigator compares the facts and analytic results to the various statements by witnesses and participants. Facts and physical evidence place limits on what actually occurred. Like a mathematical inequality or limit, they bound what is possible and what is not possible. If the statement by a witness does not fit within these limits, it should not be trusted.

For example, suppose that a particular type of subcompact car was involved in an accident. When brand new and driven by a professional driver, the car is capable of accelerating no faster than 6.5 ft/s^2 from 0 to 60 mph. It was reported by an eyewitness that the car was stopped at a red light. After the light turned green, the car then drove a short distance and passed by the eyewitness. The roadway is level. The eyewitness reported that the car flew by him and was going at least 60 mph when it passed him. Actual measurements of the distance from where the witness was standing at the time to where the car had been stopped for the red light indicate 200 ft. Should you, as the investigator, accept what the eyewitness reported?

The application of some basic physics indicates the following. Sixty miles per hour is equal to 88 ft/s. If a car accelerates at 6.5 ft/s^2, it will take 13.54 s to reach 88 ft/s. At this acceleration and lapsed time, the distance needed to accelerate to 88 ft/s is 596 ft. Consequently, the car cannot physically reach a speed of 60 mph in 200 ft. In fact, under the best circumstance, the top speed that the car can reach in 200 ft is 51 ft/s, or 34.8 mph. Because the driver was not a professional and the car had 90,000 miles on it, the actual speed that the car could attain in 200 ft was likely less than 34.8 mph. Consequently, the speed estimate reported by the eyewitness cannot be correct no matter how sincere and convincing he sounds.

Often, the analysis of the physical evidence and facts requires the simultaneous application of several scientific disciplines. In this respect, failure analysis is highly interdisciplinary. Physics, chemistry, linguistics, ethnology, art history, dentistry, engineering, microscopy, materials science, biology, psychology, dactyloscopy, sociology, psychiatry, medicine, human factors engineering, anthropology, and a host of other -ologies and arcane disciplines may be brought to bear to solve a particular failure cause problem. It may astonish some to know, for example, that entomology, the study of insects, has become a useful forensic tool in determining the time of death or the location where death occurred in certain homicide cases.

A familiarity with codes, standards, and typical work practices is also required. This includes federal, state, and local legal codes; building codes; mechanical equipment codes; fire safety codes; electrical codes; material storage specifications; product codes and specifications; manufacturer's recommendations; equipment installation methodologies; company procedures; and various safety rules, work rules, labor laws, regulations, and company policies. There are even guidelines promulgated by various organizations that recommend how some types of forensic investigations are to be conducted. Sometimes the various codes and standards, however, have conflicting requirements.

In essence, this is what an investigator does:

- He assesses the conditions that existed before the event.
- He assesses the conditions that existed after the event.
- He gathers and searches for verifiable evidence and facts, without bias or prejudice, that document and explain the transformation.

- He applies scientific knowledge and investigative skill to relate the various facts and evidence into a cohesive scenario of how the transformation occurred.
- He reports his findings and conclusions.

Implicit in the above list of what an investigator does is the application of logic. Logic provides order and coherence to all the statements, facts, principles, and methodologies that are brought to bear on a particular case. For an investigation to be first rate, the logic that connects the pieces must be sound. Methodology, no matter whose copyrighted, franchised, colorfully diagrammed, or computerized "root cause system" is being used, is not a substitute for sound logic.

The Conclusion Pyramid

In the beginning of a case, the available facts and information are like pieces of a puzzle found scattered about the floor: a piece here, a piece there, and one perhaps that has mysteriously slid under the refrigerator.

At first, the pieces are simply collected, gathered up, and placed in a heap on the table. Then, the pieces are sorted and each piece is fitted to all the other pieces until a few pieces match up with one another. When several pieces match up, a part of the picture begins to emerge. Eventually, when all the available pieces are fitted together, the puzzle is solved and the picture is plain to see.

It is for this reason that the scientific investigation and analysis of an accident, crime, catastrophic event, failure, or crime is structured like a pyramid (Figure 1.1). There should be a large foundation of verifiable facts and evidence at the bottom. These facts then form the basis for analysis according to proven scientific principles and methods. The facts and analysis, taken together, then support a small number of conclusions that form the apex of the pyramid.

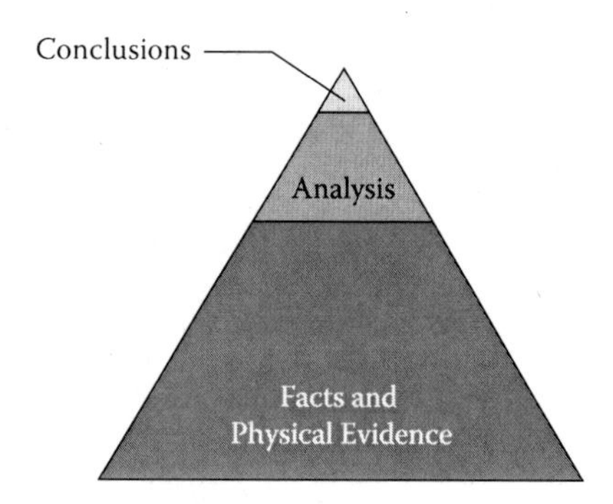

Figure 1.1 Investigation pyramid.

Conclusions should be directly based upon the facts and analysis, and not upon other conclusions, assumptions, or hypotheses. If the facts are arranged logically and systematically, the conclusions should almost be self-evident. Conclusions that are based upon other conclusions, assumptions, conjectures, or hypotheses, that in turn are based upon only a few selected facts and vague generalized principles, are a house of cards. When one point is proven wrong, the entire logical construct collapses.

Consider the following example. It is true that propane gas systems are involved in some explosions and fires. A particular house that was equipped with a propane system sustained an explosion and subsequent fire. The epicenter of the explosion, the point of greatest explosive pressure, was located in a basement room that contained a propane furnace. From this information, the investigator concludes that the explosion and fire were caused by the propane system and, in particular, the furnace.

The investigator's conclusion, however, is based upon faulty logic. There is not sufficient information to firmly conclude that the propane system was the cause of the explosion despite the fact that the basic facts and the generalized principle upon which the conclusion is based are all true.

Consider again the given facts and principles in the example rearranged in the following way:

Principle: Some propane systems cause explosions and fires.
Fact: This house had a propane system.
Fact: This house sustained a fire and explosion.
Fact: The explosion originated in the same room as a piece of equipment that used propane, namely, the furnace.
Conclusion: The explosion and fire were caused by the propane system.

The principle upon which the whole conclusion depends asserts only that *some* propane systems cause explosions, *not all* of them. In fact, the majority of propane systems are reliable and work fine without causing an explosion or fire for the lifetime of the system. Arguing from a statistical standpoint, it is more likely that a given propane system will not cause an explosion and fire.

In our example, the investigator has not yet actually checked to see if this propane system was one of the some that work fine or one of the some that cause explosions and fires. Thus, a direct connection between the general premise and the specific case at hand has not been made. *A connection has only been tacitly assumed.*

Further, not all explosions and fires are caused by propane systems; other things also cause fires and explosions. There is certainly a possibility that the explosion may have been caused by something not related to the propane system, which is unknown to the investigator at this point. The fact that the explosion originated in the same room as the furnace may simply be a coincidence, a red herring.

Using the same generalized principle and available facts, then, it can equally be concluded by the investigator, albeit also incorrectly, that the propane system did not cause the explosion. Why is this incorrect?

It is incorrect because it is equally true that some propane systems never cause explosions and fires. Since this house has a propane system, then in the same manner it could be concluded that this propane system could not have been the cause of the explosion and fire.

As is plain, our impasse in the example is due to the application of a generalized principle for which there is insufficient information to properly deduce a unique, logical conclusion. The conclusion that the propane system caused the explosion and fire is based implicitly on the conclusion that the location of the explosion epicenter and propane furnace is no coincidence. It is further based upon another conclusion that the propane system is one of the some that cause explosions and fires, and not one of the some that never cause explosions and fires. In short, in our example we have a conclusion, based upon a conclusion, based upon another conclusion.

The remedy to this dilemma is simple: Get more facts. Additional information must be gathered to either uniquely confirm it was the propane system or uniquely eliminate it as the cause of the explosion and fire.

Returning to the example, compressed air tests at the scene find that the propane piping found after the fire and explosion does not leak despite all that it has been through. Since propane piping that leaks before an explosion will not heal itself spontaneously so that it does not leak after an explosion, this test eliminates the propane piping as a potential cause.

Subsequent testing of the furnace and other appliances finds that they all work in good order also. This now puts the propane equipment in the category of the some that do not cause explosions and fires. We have now confirmed that the conclusion that assumed a cause-and-effect relationship between the location of the epicenter and the location of the propane furnace was wrong. It was simply a coincidence that the explosion occurred in the same area as the furnace.

Further checks by the investigator even show that there was no propane missing from the tank, which one would expect to occur had the propane been leaking for some time. Thus, now there is an accumulation of facts developing that show that the propane system was not involved in the explosion and fire.

Finally, a thorough check of the debris in the epicenter area finds that within the furnace room there were several open, 5-gallon metal containers of paint thinner, which the owner had presumed to be empty when he finished doing some painting work. Closer inspection of one of the containers finds that its sides are distended as if it had experienced a rapid expansion of vapors within its enclosed volume. The other paint thinner metal containers have a side that is pushed inward as if a well-distributed force had pressed upon that side of the container.

During follow-up questioning, the owner recalls that the various containers were placed only a few feet from a high-wattage light bulb, which was turned on just prior to the time of the explosion. A review of the safety labels finds that the containers held solvents that would form dangerous, explosive vapors at room temperature even when the containers appeared empty. The vapors evolve from a residual coating on the interior walls of the containers.

A back-of-the-envelope calculation finds that the amount of residual solvent in just one container would be more than enough to provide a cloud of vapor that would exceed the lower threshold of the solvent's explosion limits. Because the actual light bulb was found damaged, a check of the surface temperature of the same make and model light bulb finds that when turned on, it quickly rises to the temperature needed to ignite paint thinner fumes. A subsequent laboratory test confirms that fumes from an erstwhile empty container set the same distance away can be ignited by the same type of light bulb and cause a flash fire.

The above example demonstrates the value of the *pyramid* method of investigation. When a large base of facts and information is gathered, the conclusion almost suggests itself. When only a few facts are gathered to back up a vague, generalized premise, the investigator can steer the conclusion to nearly anything he wants. Unfortunately, there are some investigators who do the latter very adroitly.

Some Common Terms of the Art

The following are some terms usually associated with failure investigations:

- *Failure analysis* usually connotes the determination of how a specific part, component, machine, or equipment system has failed. It is usually concerned with equipment, material selection, design, product usage, methods of production, and the engineering mechanics of the failure within a component or part. However, the term is also used in connection with accidents and catastrophic events, such as building collapses or airplane crashes.
- *Evidence*, in its legal sense, is information permitted by court rules that can be put before a jury to use in resolving disputed issues of fact. This includes witness statements, documents, objects, reports, and pertinent data related to the issue. Reports concerning scientific tests, test equipment, and analyses often involve experts. The testimony of experts, however, is not limited

to the scientific tests, test equipment, or analyses contained in the report, but can also include opinions and conclusions concerning the matter that are within the purview of the expert's special knowledge, as long as the expert is obviously better qualified to assess the facts than is the jury.

There are two fundamental types of evidence: direct and circumstantial. *Direct evidence* establishes a fact that is to be proven. For example, a valid bill of sale that is not superseded by any subsequent bill of sale is direct evidence of who owns the object. Likewise, a valid birth certificate is direct evidence of a person's age.

Circumstantial evidence, on the other hand, has an element of probability associated with it. The fact to be proven must be inferred by at least one logical step. For example, consider that John and Mary have been observed to live together by neighbors for many years. They raised a family and have led a typical married lifestyle. These observations might be used to infer that John and Mary were legally married. They act married; therefore, it is concluded that they are legally married.

According to the *2005 Time Almanac* (p. 127), of the coupled households in the United States in 2000, 90.9% were married partners and 9.1% were unmarried partners. Thus, the conclusion that John and Mary are legally married has a 90.9% chance of being correct, if no other factors are known. Of course, this also leaves a 9.1% chance that John and Mary are not legally married, that they are just living together. This then begs the question: What is risked if the conclusion that they are legally married is made, but it turns out to be wrong?

With respect to courts, with some exceptions, evidence is admitted to court proceedings if it is relevant to the matter in dispute. It must, in some way, either prove or disprove a question of fact. Direct evidence is usually admitted. Circumstantial evidence may be admitted, depending upon the judge's discretion, legal precedent, or special rules. The four common reasons for evidence being denied admittance to a court proceeding are:

- When the evidence is hearsay
- When it is ruled irrelevant
- When it is legally privileged information
- When the evidence was obtained by illegal means

- *Root cause analysis* (RCA) is often used in connection with the managerial or human performance aspects of failures rather than the failure of a specific part, and how procedures and managerial techniques can be improved to prevent the problem from reoccurring. Root cause analysis is often used in association with large systems, such as power plants, construction projects, and manufacturing facilities, where there is a heavy emphasis on safety and quality assurance through formalized procedures. It is generally presumed that if the true root cause is identified and then fixed, the event will not recur, at least not in the same way.

- The modifier *forensic* in *forensic engineering*, *forensic investigation*, and *forensic science* typically connotes that something about the investigation of the event will relate to the law, the courts, adversarial debate, or public debate and disclosure. Forensic engineering or forensic science can be either specific in scope, such as failure analysis, or general in scope, such as root cause analysis. It all depends upon the nature of the dispute.
- *Contributing cause* denotes a factor that increases the severity of the resulting event, increases the likelihood of its occurrence, or causes the event to be advanced in time or extent. However, by itself, a contributing cause cannot cause the event to occur. An example of a contributing cause is drums of gasoline stored in an area where there is electrical sparking. While the sparking may initiate a fire, the drums of gasoline increase the chances that any resulting fire will be severe and extensive. The drums of gasoline, however, cannot start the fire.
- A *causative factor* is an action, state, or condition that existed before the event that increased the likelihood that the event would occur. There are two types of causative factors: an indirect cause and a direct cause.
- A *direct cause* is an action, state, or condition that existed immediately before the event occurred and directly allowed or promoted the occurrence of the event.
- An *indirect cause* is the same thing as a contributing cause. It sets the stage, increases the probability, or increases the severity of an event. An indirect cause, however, cannot by itself cause the event to occur.
- A *root cause* is a type of direct cause. It has been defined by some as the fundamental cause of the event, which when removed, modified, or corrected, would have prevented the event from occurring in the first place. Alternately, it is the fundamental cause, which when removed, modified, or corrected, will prevent the event from recurring.

The preceding definition tacitly implies, however, that a failure has only one root cause. For a time, some root cause instructors did teach this tenet. They taught that all failure events could eventually be traced back to a single root cause, and that if a person had found more than one root cause, then the investigation had simply not gone deep enough. In that theory, there was to be one and only one root cause. This "one event–one cause" tenet had a strong philosophical appeal to some.

However, few experienced failure investigators agree that all failures, that is, all cause-and-effect relationships, have one and only one root cause. Some events do, of course, have a single root cause. However, even those who once were adherents of one event–one cause have largely abandoned the tenet.

The current working definition of *root cause* is a fundamental cause that leads to the failure event either by itself or in concert with other root causes. Further, it is a cause that can be modified or affected so that another like failure can be stopped, prevented, mitigated, or caused to occur less frequently. Consequently in this pragmatic definition, a cause over which a person has no control cannot be a root cause.

An example of this is lightning. Consider lightning damage to a switchyard. While it might be said that lightning caused damage to equipment in the switchyard, people do not have control over lightning as to when and where it might occur. However, people do have control of how lightning is handled in a transmission line by means of lightning protection such as arrestors, appropriate grounding, and so on. Thus, while lightning itself may cause damage, the root cause may involve a lack of appropriate equipment to safely handle the lightning.

- An *apparent cause* is typically the cause of an event as determined by a limited investigation. This often means that the evidence already at hand is considered, but a detailed, methodical investigation is not undertaken. An apparent cause investigation may include some assumptions based upon experience or probability that may not be explicitly verified. Because an apparent cause investigation is less rigorous than a root cause investigation, the risk of reaching an inaccurate conclusion is higher. However, the cost and amount of time involved with an apparent cause investigation are less. Thus, an apparent cause investigation is often undertaken when the problem is small or limited in scope, and the risk or consequences of being wrong can be tolerated.
- A *programmatic cause* is a deficiency in a procedure, program, training lesson, or other managerial construct that contains a step or instruction that increases the likelihood that a human error will occur. An example is the wording in an instruction that has several interpretations. Another example is a complex procedure that is required to be executed at a time when the worker is fatigued, in an uncomfortable environment, or under pressure to do the work quickly.
- A *reconstruction* is when a particular failure, crime, or event has been explained: how it came about, the sequence of events, the important factors, who or what was involved, what items caused which events to occur, and so on.

- *Human performance evaluation process* (HPEP) is the method used to assess how people's actions contributed to the causation of the event. For example, if an event depended upon a person's actions to be successful, and instead the event was a failure:
  - Did someone fail to act?
  - Did someone act but make the wrong choice?
  - Was the person fatigued, stressed, or error-prone?
  - Was the person properly trained to act correctly?
  - Did the person have the knowledge to make the correct decision?
  - Was the person distracted or otherwise prevented from acting correctly?
  - Did the person have a motive to promote the failure by increasing its likelihood, allow the failure to occur by inaction, or directly cause the failure?
  - Were there motive, means, and opportunity if the failure was deliberate?
  - Was the person physically capable to act properly?
  - Did the person have the correct tools or gear?
- After the fundamental cause for an event has been determined, a *corrective action* is what is done to remedy the weakness or problem identified in the root cause. The intention of the corrective action is to prevent the event from recurring due to the same reason, to mitigate the consequences if it does recur due to the same reason, or to reduce the likelihood that the event will recur due to the same reason.
- After an investigation is done and the root cause and contributing causes have been determined, the *extent of condition* is sometimes evaluated. An extent of condition evaluation answers the following questions: Can this same type of failure or event occur elsewhere? Do the same conditions discovered in this case exist elsewhere? The idea is to use the knowledge gained from the investigation just completed to prevent similar events from occurring again elsewhere.

  A simple example is a faulty switch. If a failure were caused by a specific model switch that was found to have a latent manufacturing defect, other future failures could be averted if all the same model switches were found and replaced with good ones.
- *Falsification* is a principle used in the application of the scientific method. The verification of a working hypothesis not only requires that the hypothesis provide some predictive value, that is, be testable, but also requires that the hypothesis not be proven incorrect. Facts and data that could prove that a hypothesis is wrong must be as vigorously pursued as facts and data that appear to prove a hypothesis correct.

Crime versus Failure

What is the difference between the investigation of a crime and the investigation of a failure, accident, or catastrophic event? To answer that, first let us examine the differences in their definitions.

- A *crime* is a willful action, a failure to act, or a negligence of duty that is contrary to public law or is specifically prohibited by law. Those who are found guilty of a crime are punished in accordance with penalties prescribed by law. Occasionally, they are required to make restitution.
- An *accident* is an unintentional, undesirable occurrence. Often it is considered to be the result of bad fortune, ill chance, destiny, the fates, happenstance, or unfortunate coincidence. Usually, measures are taken after an accident to prevent it from recurring or to mitigate its consequences if it should occur again. An accident can also be a crime.
- An *act of God* is a direct and irresistible action of nature or turn of events involving natural forces, such as storms, floods, tsunamis, lightning, droughts, and earthquakes. Typically, the term *act of God* describes a catastrophic event that cannot have been caused by a person. However, acts of God can sometimes be foreseen, avoided, or mitigated by people. This is the basis for earthquake provisions in building codes, flood control systems, and weather warning systems. Some acts of God are sometimes also considered to be accidents. The failure to provide for the mitigation of acts of God can sometimes be a crime, e.g., not providing tornado or hurricane safety shelters in mobile home parks, or not providing a neighborhood water hydrant system to put out fires caused by lightning.
- A *catastrophe* is a sudden disaster or fiasco that is relatively widespread. An act of God such as a flood can be a catastrophe. An accident, such as loss of electrical power over a large region during a heat wave can be a catastrophe. A catastrophe can also be a crime, for example, the sabotage of the World Trade Center on September 11, 2001.
- A *failure* is a nonperformance, insufficiency of performance, or performance below expectations. Failure can also be the degradation, decay, or deterioration of performance or material quality. Failures can also be accidents, catastrophes, acts of God, and sometimes crimes.

The one common characteristic among all the above terms is that the resulting event is undesirable. In the initial stages of an investigation, when knowing only something about the resulting event, it is not readily apparent that the cause was due to a crime, an accident, or a failure. In fact, if the accident or failure was caused by the misjudgment of a person, the only difference between an accident, failure, or crime is perhaps the condition of willfulness, the motive, or whether the situation is addressed in the legal code.

In short, crimes, accidents, acts of God, and failures are the results of cause-and-effect phenomena. Fundamentally, they are investigated in the same way to determine how they came about.

There is an important distinction to note between crimes and failures that involve prosecutable liability, and other kinds of failures and accidents that may not involve prosecutable liability. The law prescribes what kind of evidence is required to convict a person of a crime or find him liable for damages. Further, the law also prescribes what kind of evidence is acceptable to the court, what weight it is to be given, and sometimes how it is to be obtained or not obtained.

To ensure that the facts and evidence being gathered remain court-worthy, care must be exercised to ensure that the evidence has been obtained, transported, handled, analyzed, and preserved in accordance with practices accepted by the court. Consequently, evidence to determine if a company procedure is deficient may be gathered to less stringent standards than evidence to convict a person of murder. This is to be expected. When the consequences of an investigation can result in the loss of fortune, dishonor, incarceration, or capital punishment, high standards of care are appropriate.

How Accidents and Failures Occur

Often, an accident or failure is not the result of a single cause or event. It may be the combination of several causes or events acting in concert or in sequence. An example of causes acting in sequence might be a gas explosion.

- A spark from a pilot light ignites accumulated gas.
- The gas originated from a leak in a corroded pipe.
- The pipe corroded because it was poorly maintained.
- The poor maintenance resulted from an inadequate maintenance budget such that other items were given higher priority.

An example of causes acting in concert might be an automobile accident.

- It is dusk.
- Driver A has to yield to approaching traffic to make a left turn while the light is green. He doesn't signal his left turn, and while the light is green there is too much traffic for him to turn. At the end of the green light, he suddenly turns left, assuming there will be a gap in the traffic after the last car has passed and the light has turned yellow.

- Coming from the opposite direction, driver B enters the intersection at the tail end of the yellow light. He guns the accelerator so that he will get across the intersection while the light is still yellow.
- They collide in the middle of the intersection. Car B impacts car A on the passenger side, just before car A could complete his left turn.
- Driver A is drunk. His reflexes are slow and his judgment is impaired.
- Driver B's car has bad brakes that do not operate well during hard braking. Driver B is also driving without his glasses, which he needs to see objects at a distance, especially at dusk.

Often, failures, accidents, and crimes involve both sequential events and events acting in concert in various combinations. One of the more important strategies for preventing failures is the interruption of this chain of events. Sometimes changing the outcome of just one sequential or parallel event can prevent the failure from occurring.

Eyewitness Information

As was briefly discussed previously, eyewitness accounts are important sources of information, but they must be carefully scrutinized and evaluated. Sometimes eyewitnesses form their own opinions and conclusions about what occurred. They may then intertwine these conclusions and opinions into their account of what they say they observed. Skillful questioning of the eyewitness can sometimes separate the factual observations from the personal presumptions.

Consider the following example. An eyewitness initially reports seeing "Bill" leave the building just before the fire broke out. However, careful questioning reveals that the eyewitness did not actually see Bill leave the building at all. The witness simply saw someone drive away from the building in a car similar to Bill's. The witness simply concluded on his own that it must have been Bill. Of course, the person driving the car could have been Bill, but it also could have been someone with a car like Bill's, or someone who had borrowed or stolen Bill's car. Bill was not actually verified by direct observation as being the driver.

Of course, some eyewitnesses are not impartial. They may be relatives, friends, or enemies of persons involved in the event. They may have a personal stake in the outcome of the investigation.

For example, it is not so unusual for the arsonist who set the fire to be interviewed as an eyewitness to the fire. Arsonists sometimes hang around to watch their own handiwork being put out. Let us also not forget those eyewitnesses who are simply scalawags, psychotics, and blowhards. They may swear to anything to pursue their own agendas or get attention.

What an honest and otherwise impartial eyewitness reports observing may also be a function of his location with respect to the event. His perceptions of the event may also be colored by his education and training, his life experiences, his physical condition with respect to eyesight or hearing, and any social or cultural biases. For example, the sound of a gas explosion might variously be reported as a sonic boom, cannon fire, blasting work, or an exploding skyrocket. Because of this, eyewitnesses to the same event may sometimes disagree on the most fundamental facts.

Further, the suggestibility of the eyewitness in response to questions is also an important factor. Consider the following two exchanges during *statementizing.*

Exchange I

Interviewer: Did you hear a gas explosion last night at about 3 a.m.?
Witness: Yeah, that's what I heard. I heard a gas explosion. It did occur at 3 a.m.

Exchange II

Interviewer: Tell me what happened last night.
Witness: Something loud woke me up.
Interviewer: What was it?
Witness: I don't know. I was asleep at the time. It was loud, though.
Interviewer: What time did you hear it?
Witness: I don't know exactly. It was sometime in the middle of the night. I went right back to sleep afterwards.

In the first exchange, the interviewer suggested the answer to his question. Since the implied answer seems logical, and since the witness may presume that the interviewer knows more about the event than himself, the witness agrees to the suggested answer. Afterwards, the witness may repeat the information suggested to him by the interviewer as if it were his own original recollections.

In the second exchange, the interviewer did not provide any clues to what he was looking for. He allowed the witness to draw upon his own memories and did not suggest any. This is preferred.

Some Investigative Methods

The following are some of the methods commonly used in the investigation of failures, accidents, crimes, and similar events. Each will be discussed in subsequent chapters.

- Scientific method
- Fault tree analysis
- Why staircase
- Event and causal factors analysis (ECFA)
- Barrier analysis
- Human performance evaluation process (HPEP)
- Change analysis
- Motive, means, and opportunity analysis
- Organizational and programmatic analysis
- Management organizational risk tree (MORT)
- Kepner-Tregoe, Apollo, Tap Root, and similar proprietary methods

Role in the Legal System

From time to time, a person who does this type of investigative work is called upon to testify in deposition or in court about the specifics of his or her findings. Normally the testimony consists of answers to questions posed by an attorney for an involved party. The attorney will often be interested in the following:

- The investigator's qualifications to do this type of analysis
- The basic facts and assumptions relied upon by the investigator
- The reasonableness of the investigator's conclusions
- Plausible alternative explanations for the accident or failure not considered by the investigator, which often will be his client's version of the event

By virtue of appropriate education and experience, a person may be qualified as an expert witness by the court. In some states, such an expert witness is the only person allowed to render an opinion to the court during proceedings. Because the U.S. legal system is adversarial, each attorney will attempt to elicit from the expert witness testimony to either benefit his client or disparage his adversary's client.

In such a role, despite the fact that one of the attorneys may be paying the expert's fee, or the expert may be in the employment of one of the parties involved in the proceedings, the expert witness has an obligation to the court to be as objective as possible and to refrain from being an advocate. The best rule to follow is to be honest and professional both in preparing the original analysis and in testifying. Prior to giving testimony, however, the expert witness has an obligation to fully discuss with his or her client both the favorable and unfavorable aspects of the analysis.

Sometimes the investigator involved in preparing an accident or failure analysis is requested to review the report of analysis of the same event by the expert witness for the other side. This should also be done honestly and professionally. Petty one-upmanship concerning academic qualifications, personal attacks, and unfounded criticisms is unproductive and can be embarrassing to the person who engages in it. When preparing a criticism of someone else's work, consider what it would sound like when read to a jury in open court.

Honest disagreements among qualified experts can and do occur. Sometimes the basis for the disagreement is the set of facts each has collected to use for analysis of the event. Each investigator may have a different set of facts from which to work. Sometimes the basis for disagreement is missing facts, that is, gaps in the story or the timeline, and how the investigator logically bridges those gaps. In any case, when such disagreements occur, the focus of the criticism should be the theoretical or factual basis for the differences, not *ad hominem* attacks.

The Fundamental Method 2

> The human mind is prone to suppose the existence of more order and regularity in the world than it finds.
>
> **Sir Francis Bacon, 1620**

The Fundamental Basis for Investigation

The fundamental basis for failure investigations is the scientific method.

As noted in the standard concerning the investigation of a fire or explosion published by both the American National Standards Institute and the Nation Fire Protection Association, ANSI/NFPA 921, Chapter 2, 1992 edition:

> A fire or explosion investigation is a complex endeavor involving both art and science. The compilation of factual data, as well as an analysis of those facts, should be accomplished objectively and truthfully. The basic methodology of the fire investigation should rely upon the use of a systematic approach and attention to all relevant details. The use of a systematic approach often will uncover new factual data for analysis, which may require previous conclusions to be re-evaluated.
>
> The systematic approach recommended is that of the scientific method, which is used in the physical sciences. This method provides for the organizational and analytical process so desirable and necessary in a successful fire investigation.

Likewise, the *Guidelines for Failure Investigation*, published by the Task Committee on Guidelines for Failure Investigation of the American Society of Civil Engineers in 1989, recommends utilization of the scientific method, which is described on p. 62, Section 2.12:

> The process of developing failure hypotheses varies with the complexity of the failure situation. During an investigation of a minor nature the forensic engineer might develop failure hypotheses working alone. However, during the investigation of a large-scale failure requiring a team of investigators, the development of failure hypotheses should involve discussions and conferences during several phases of the investigation plan. Thus, the development of a final failure hypothesis should be an evolutionary process with input from the entire investigative team.

> Often after the initial site visit, a failure hypothesis may be suggested; then as data is collected and analyzed, the initial hypothesis may be proven, abandoned, or revised, based on the synthesis of investigative data. If the initial hypothesis is abandoned and a new hypothesis is generated, then additional site investigation, document review, and synthesis of the analysis process may be required until a final hypothesis is established.

The above is true not only for fire, explosion, and civil engineering failure investigations, but also for the investigation of other types of failures, crimes, and significant events whose causes need to be determined. Other nationally recognized investigative codes and standards have similar statements that endorse, and in some cases require, the use of the scientific method.

The Scientific Method

There are historically two versions of the basic scientific method. The first and older version involves collecting verifiable facts and observations about an effect, event, item, or subject. A person then assesses these facts and observations and posits a general proposition consistent with the data. As new facts and observations accumulate, the general proposition may be modified or changed to maintain consistency with the known body of verifiable information.

For example, suppose data concerning the wingspan, weight, and flying ability of various bird species were collected and correlated. Based upon these correlations, generalizations about the ratio of wingspan to weight that is necessary to successfully achieve long-distance flight, medium-distance flight, or short-distance flight, or perhaps not being able to fly at all, might be concluded.

This was the basic method, which is sometimes called empiricism, advocated by Lord Francis Bacon who lived in the late sixteenth and early seventeenth centuries. Lord Bacon, by the way, is not to be confused with Friar Roger Bacon, who also had a hand in developing the scientific method several centuries before Lord Francis Bacon. The role each played in developing the scientific method will be noted in the next chapter.

Fundamentally, empiricism is the tenet that knowledge is derived from measurements, verifiable facts, observations, and actual, as opposed to spiritual or supernatural, experiences, feelings, or beliefs. Furthermore, as they are collected, these measurements, verifiable facts, observations, and experiences are to be reported and documented without being modified or interpreted by preconceived notions.

For example, if the ignition temperature of a certain kind of paper is reported in reference books to be 451°F, this fact can be verified by anyone, in any laboratory, by recreating the same conditions as specified in the reference and measuring the temperature when ignition occurs. This is an example of what is meant by the terms *repeatable* and *verifiable*.

If an investigator, however, finds that three samples of the same kind of paper listed in the reference book ignite variously at 454°F, 450°F, and 447°F, after ensuring that the specified conditions have been faithfully recreated and this is not due to instrumentation variances, then these are the ignition temperatures that should be reported.

If our dauntless investigator, however, has a preconceived notion that the correct answer must be 451°F, since this is the value listed in the reference book, then the investigator might be tempted to report that he found that the ignition point was indeed 451°F, and to ignore the small differences he actually observed. The investigator might presume that he must have done something slightly wrong to have missed, but come close to, the correct answer. After all, 454°F, 450°F, and 447°F are all close to 451°F, and taken together, the three values do have an average of 451°F.

The outcome of this is that the investigator nudges the data to fit his preconceived notion of what he expected to find. This is called a confirmation bias. In doing so, he fails to report what was plainly in front of his eyes: that the kind of paper he tested has small but measurable variations in ignition temperature from that reported in the reference book.

This variation in ignition temperature may become important to the investigation in ways the investigator cannot imagine at this point. Unfortunately, in his eagerness to confirm the presumed correct value listed in the reference book and not appear wrong, the investigator perhaps threw away a very useful nugget of knowledge and compromised sound laboratory practice. A good rule for field or laboratory work is to report what is actually observed, not what a person thinks should be observed or what he has concluded he has observed.

In this regard, consider the following story researched and reported by Lisa Jardine in the April 28, 2006, online BBC News. In 1664, sea trials were done by the British Navy to test the use of pendulum clocks to determine longitude. The clocks were taken on a 9-month voyage southward down the west coast of Africa. Upon their return to London in 1665, the captain of the expedition, Captain Robert Holmes, reported that the clocks had worked so well, they had actually saved the lives of the crews of the four ships.

What was reported by Captain Holmes was this: On the return journey, Captain Holmes and the expedition had to sail several hundred miles westward in order to catch a favorable wind. In doing so, all four ships had run low on drinking water—so low, in fact, that they thought they were in danger of running out of water. According to their traditional calculations, the expedition was far from any freshwater supply point.

Captain Holmes, however, reported that according to the pendulum clocks they were only a short distance west of Fuego, an island in the Cape

Verde group. The expedition sailed due east accordingly and made landfall at Fuego the next day, as predicted. When news reached London about the success of the pendulum clocks in accurately determining longitude, the story was heralded about how well the clocks had worked. The Fellows of the Royal Society were impressed, and orders for the new clocks came in.

However, the inventor of the clock, Christiaan Huygens, had doubts about this claim. He did not believe that his own invention could be so accurate. In fact, Mr. Huygens doubted that the captain honestly reported the event. Ms. Jardine reports that Mr. Huygens said the following in a letter to the Royal Society:

> I have to confess that I had not expected such a spectacular result from these clocks. I beg you to tell me if the said Captain seems a sincere man whom one can absolutely trust. For it must be said that I am amazed that the clocks were sufficiently accurate to allow him by their means to locate such a tiny island.

Consequently, the Royal Society asked Mr. Samuel Pepys to check the evidence supporting the claim, especially the ship log of Captain Holmes. Pepys's scrutiny of the matter found cause for Mr. Huygens's doubts. The pendulum clocks had not been any more accurate than the traditional calculations by the other captains in the expedition. The sailors had simply been lucky. Captain Holmes falsified the evidence in order to please the fellows at the Royal Society. In other words, Captain Holmes told them what they wanted to hear and what the evidence appeared to support.

It is commendable that Mr. Huygens was scrupulously logical. Even though he had invented them, he knew that his pendulum clocks could not be as accurate as was being claimed. In fact, Mr. Huygens's pendulum type ship clocks never were particularly valuable in accurately determining longitude. An effective ship's clock to determine longitude was not invented until a hundred years later, in the mid-1700s, by John Harrison. Thus, the moral of this story is that even when faced with agreeable evidence, it is important to have a questioning attitude. How many sailors might have died by fatal navigation errors if the clocks had been presumed to be accurate?

Empiricism is contrasted by rationalism. Rationalism is the tenet that knowledge can be derived from the power of intellect alone. In other words, knowledge can be derived by deductive reasoning from basic principles that people sometimes take for granted without physical proof or justifying evidence.

Sometimes these basic principles are said to be self-evident; that is, they are considered to be so obvious that no proof is required. Alternately, they are described as being self-evident because the principles are so fundamental, their proof would constitute a tautology; that is, they can only be proven by referring to themselves. Notions about beauty, symmetry, religion, ideology, truth, good and evil, and philosophy are often used to justify self-evident principles.

On a practical level, most notions about beauty, symmetry, and so on do appear to change with time and vary from culture to culture. Despite protestations to the contrary from "true believers," notions about truth, right and wrong, beauty, symmetry, and so on are not held absolute even within the same culture. In literate cultures, the evolution of principles held to be absolute is well documented in the historical record. In other cultures, it is indirectly documented through their art, mythology, and traditions.

After 2,000 years of acceptance, for example, the self-evident principles of geometry put forward by Euclid were eventually challenged by nineteenth-century mathematicians as not really being self-evident. The result was non-Euclidean geometry, which was later used to describe the time–space relationships embedded in relativity. Euclid's geometry was simply one kind of geometry among many.

Unlike empiricism, rationalism allows preconceived notions to provide a framework into which new facts and data are fitted. Some fundamental principle is usually considered inviolate; that is, the principle can't be changed or modified because it is assumed to be absolutely true. Consequently, in order for the fundamental principle to remain inviolate, all data, facts, and observations must be made to be consistent with the fundamental principle.

Data that appear to contradict the assumed fundamental principle have to be discredited or must be explained away by the invention of new rules. Sometimes the new rules apply only to that particular fact or group of facts. In the preceding example involving the ignition temperature of paper, the investigator did just that: He discredited his own experimental data when it did not agree with the value he expected. He presumed he had made some small mistake that caused the variation and then nudged the figures to make his laboratory results agree with the value listed in the reference book.

Consider the following simple example about disease to highlight the difference between empiricism and rationalism. Empiricists might note the statistical distribution of a disease, the various age groups it affects and their occupations, when it occurs, where it occurs, and so on. In this way perhaps a useful correlation about the disease may be learned that allows the disease to be avoided, prevented, predicted, or perhaps cured. This was how, for example, the connection between yellow fever and mosquitoes was made.

Rationalists, on the other hand, might propose as a fundamental principle that disease is a malady sent by God to punish people for wrongdoing. Consequently, even if it appears that good people are being stricken by disease, which would contradict the basic principle if taken at face value, rationalists might conclude that disease is striking down otherwise good people because they have done bad things in secret and only God knows their secrets. Disease, therefore, provides an excellent way for mortals to distinguish between people who are guilty of sins committed in secret and people who are genuinely good.

Thus, the observation that seemingly good people are also struck down by the disease is explained away by the invention of a special rule. Note that the new rule, like the fundamental principle, does not lend itself to being either proven or disproven. It is a matter of faith, not evidence. Further, because God is presumed to be the controlling factor with respect to disease, some people might conclude that there is little point in attempting to do anything to understand or avoid disease. Mere mortals should not interfere with God's will.

This was the basic premise of the famous, or more honestly the infamous, published sermon of Reverend Edward Massey in 1772, *The Dangerous and Sinful Practice of Inoculation.* As a result of opposition to inoculation against smallpox by various religious leaders, thousands in Europe and in the American colonies died who could have otherwise been saved from the scourge. Further, despite the reported successes of the Jenner method of vaccination, which removed the risks associated with the live inoculation method, in 1798 various clergymen in Boston formed the Anti-Vaccination Society. This society called upon people to refuse vaccination because it was "bidding defiance to Heaven itself, even to the will of God," and that "the law of God prohibits the practice." Jenner was still being regularly vilified from the pulpit at the University of Cambridge in 1803.

Lest some think that the preceding notion about disease is absurd or just an academic discussion of a belief no longer harbored by modern civilization, please scan the Internet under topics such as *disease and punishment or disease as punishment by God.*

In short, empiricists accept what is observed at face value and try to find order to explain it. Rationalists, on the other hand, propose up front what the order should be and then sort their observations and facts to fit within that framework. More will be said about this later when *a priori* and *a posteriori* reasoning are discussed.

The second and more modern version of the basic scientific method also involves collecting data about a particular effect and looking for a principle common to the observations. After developing a working hypothesis consistent with the available data, however, the hypothesis is then applied to anticipate additional consequences or effects that have not yet been observed. In short, the hypothesis is tested to determine if it has predictive value.

These additional consequences or effects are then sought out: perhaps by experimentation, perhaps by additional observations, or perhaps by reexamining evidence and data already collected. Often, a good working hypothesis allows a person to predict the presence of evidence or effects that were present from the beginning but were overlooked, misunderstood, or considered unimportant during the initial evidence gathering period.

If the additional consequences or effects predicted by the hypothesis are confirmed, this does not automatically verify the hypothesis, however. It is only a tentative confirmation of the hypothesis. When a prediction and its subsequent verification apparently support the hypothesis, do not contradict it, and demonstrate that the hypothesis has useful predictive ability, it still doesn't conclusively prove the hypothesis.

Verification of a hypothesis occurs when observations, or experiments and data that could show the hypothesis to be false fail to do so. This is the principle of falsification. Thus, there must be not only verifiable evidence that supports a hypothesis, but also no verifiable evidence that falsifies the hypothesis.

The last point is important and is often the most forgotten or neglected step in the scientific method. A reasonable and unbiased effort must be expended to discredit the hypothesis. Verifiable evidence that appears to falsify a hypothesis should not be overlooked, neglected, or dismissed. If a hypothesis truthfully and faithfully explains the phenomenon or event that occurred, it will stand against tough, critical scrutiny. Conclusions based upon inductive reasoning alone may be flawed unless they are well tested by falsification.

In sum, the attributes of a good working hypothesis are:

- All the data upon which it is based needs to be factually verifiable.
- It must be consistent with all the relevant verifiable data, not just selected data.
- The scientific principles upon which the hypothesis relies must be verifiable and repeatable.
- The hypothesis should provide some predictive value.
- The hypothesis must be subjected to and withstand genuine falsification efforts.

The Value of Falsification

Consider the hypothesis that the earth is flat. After making hundreds of personal casual observations from his personal vantage point unaided by any instrumentation, an investigator concludes that the earth is flat. Certainly people living in south-central Illinois, the Pampas in Argentina, the Sahara Desert, or the Panhandle of Texas might easily concur that the earth looks flat to them, too, as they go about their daily routine.

At the beach, the investigator further observes that the horizon dividing the earth from the sky appears to be a straight line, which he concludes also supports his flat earth hypothesis. A search of historical records by the investigator also finds ample precedent that famous and important people from times past also believed that the earth is flat. (This is called validation by appeal to authority.)

Based upon the flat earth hypothesis, the investigator predicts that the Outback of Australia will appear flat and that the Mongolian Desert will also appear flat. After traveling there, by surface travel, of course, these predictions appear to be born out: both places look flat to the investigator. In fact, everywhere the investigator looks with his unaided eyes, the earth appears flat. Eventually, the investigator collects an overwhelming pile of anecdotal evidence that seemingly proves that the earth is flat.

But, of course, the earth is not flat. It only appears flat when observed in sufficiently small pieces with unaided eyes. A single, panoramic observation from space or a single circumferential navigation of the earth, as was done by Magellan, supersedes any mountain of accumulated anecdotal evidence that supports the flat earth hypothesis. Thus, it is not the quantity of evidence that supports a hypothesis that counts, or the perceived authority of people who agree with the hypothesis that counts. What counts is the quality of the facts that falsify a hypothesis, or perhaps fail to falsify a hypothesis.

When a person tends to notice only evidence that supports a certain hypothesis and undervalues, dismisses, or doesn't even bother to look for contradictory evidence, this is a confirmation bias. Gamblers are famous for doing this. Gamblers tend to remember best the times when they win and tend to forget or minimize the memories of the times when they lose. Surveys of typical gamblers leaving casinos show that most believe they have won more than they have lost, or at least that they have nearly broken even, when the facts plainly indicate that more people lose money in casinos than win money.

Investigators sometimes become caught up in confirmation bias traps of their own making. They may set up laboratory tests, recreations of the event, or conduct other sorts of tests whose outcomes are framed to support a certain hypothesis. Concomitantly, they may avoid dealing with data or outcomes that might contradict the hypothesis in question. This is especially true when much investigative time and energy has already been invested in a certain hypothesis, and deadlines and budgets predispose the investigator against initiating a major new line of inquiry.

The Michelson–Morley experiment, wherein the presence of cosmic aether was not detected, is a famous example of how falsification works. For many years, prominent scientists had proposed the presence of a cosmic substance, aether, which supposedly permeated the universe. Aether existed, they argued, to explain how light is transmitted from the sun to the earth or from the stars to the earth. Since these scientists assumed that light propagated as a wave, like sound in air or a vibration though solid material, they concluded there must be some type of substance permeating space that provides a medium for the transmission of light. The aether existed, they argued, because there was just no other way to account for the transmission of light through empty space.

By the way, the expression "there is just no other way" and similar expressions, such as "it can't be anything else," "it is impossible for anyone else but...," and "by elimination it must be ...," are indications that a thinly disguised argument by exhaustion is being employed. Be wary of such arguments. While it is certainly a *bona fide* logical method when employed properly, it is also one of the most abused logical methods and is a great argumentation refuge for charlatans and pseudointellectual rogues. Almost anything can be and has been proven by this method when the underlying assumptions are not made plain and carefully examined. The key is understanding the basis and limitations of the exhaustive list from which items have been eliminated until only one possibility remains.

Consequently, if light propagates through the cosmic aether like sound waves traveling through air or water, light should be measurably faster in the direction of travel and should be measurably slower in a direction perpendicular to or opposite to that of travel. This concept is analogous to sound emitted by a person riding a bus.

When a person riding a bus makes a sound that emanates equally in all directions, another person standing still on the ground and in front of the bus will measure the sound to be slightly faster in the direction of travel of the bus.

Sound velocity + bus velocity = slightly faster sound velocity

Likewise, the nonmoving observer will note that sound heard when the observer is standing perpendicular to the direction of travel of the bus will be unaffected by the motion of the bus.

Sound velocity + 0 = same sound velocity

The sound velocity measured by the observer standing behind the bus is then slightly slower in the direction opposite the travel of the bus.

Sound velocity – bus velocity = slightly slower sound velocity

Based upon these arguments, Michelson and Morley in 1887 attempted to measure the speed of the earth relative to the aether. Presuming that the aether theory was true, it should be possible to measure the speed of the earth through the aether by comparing the length a beam of light travels in a certain period of time in the direction of travel of the earth to a similar beam in the same amount of time directed perpendicular to the earth's direction, or directed in the opposite direction of travel. A comparison of the lengths of the two light beams, and the application of a little algebra, would then allow the speed of the earth relative to the aether to be determined.

Using a sensitive interferometer, a device that compares one length of light against another, the two experimenters found that light traveled at the same speed in the direction that the earth was traveling as it did either in a direction perpendicular to the motion of the earth or in the opposite direction to the earth's motion. In fact, despite the fact that they knew the earth was moving around the sun with respect to the solar system at 66,700 mph, they found that light had measurably the same speed no matter which direction from earth they looked. In short, to them the speed of light appeared constant in all directions despite the obvious motion of the earth.

Initially, Michelson and Morley thought they had done something wrong. Consequently, they repeated their experiment many times and thoroughly checked their equipment. After exhaustively ensuring that the fault was neither the method nor the equipment, they published their results.

The null result of the Michelson–Morley experiment eventually led to the development of the theory of relativity by Albert Einstein. Some years later, physicists discovered that they could account for the transmission of light through empty space without a transmission medium that permeated all of space. Their previously held assumption that there was simply no other way to account for the transmission of light through space was wrong because their list of possibilities was not truly exhaustive. They had omitted something that they didn't know at the time: Electromagnetic waves, including light, can propagate through a vacuum due to the self-propagating nature of alternating electric and magnetic fields. For this work Michelson was the first U.S. citizen awarded the Nobel Prize for physics.

Consider how the principle of falsification made a difference to death row inmates in Illinois. On January 31, 2000, Governor George Ryan announced a moratorium on executions because Illinois had recently exonerated more death row inmates than it had executed.

Since executions had resumed in 1977 in Illinois, thirteen inmates had been exonerated while twelve had been executed. A fourteen-member commission was charged to investigate how thirteen innocent persons were waiting to be executed despite safeguards in the legal system to prevent false capital crime convictions. Governor Ryan stated, "Until I can be sure with moral certainly that no innocent man or woman is facing lethal injection, no one will meet that fate."

What had happened? When the thirteen people in question had been tried for capital offenses, DNA testing was not available. With the available evidence at the time of trial, some of which even included blood typing and eyewitness testimony, these thirteen persons were convicted and sentenced to death by juries of their peers. The evidence presented at the time of trial was apparently persuasive and left no doubt in the jurors' minds. These cases were then reviewed and appealed according to the standard legal procedures in Illinois to ensure that the trials were fair and impartial. Despite these safeguards, however, these thirteen people were waiting their turn to die.

Fortunately, when DNA testing became available for routine forensic work, some of the stored evidence in these thirteen cases still had usable bloodstains, semen, or other genetic material from the culprit that could be tested and compared with that of the person convicted of the crime. DNA testing, when properly done, is verifiable and repeatable.

In these thirteen cases the results of DNA testing excluded the person convicted of the crime from being the culprit. In other words, the DNA tests falsified the evidence previously provided to the court that supported a guilty verdict.

Similar events have unfolded in nearly all U.S. states since about 2000, when DNA testing began to be used to aggressively check previous guilty verdicts that were primarily based on traditional eyewitness identification testimony and circumstantial evidence. Since 2001, for example, the Dallas, Texas, courts have had to release sixteen people (as of May 26, 2008) who had been previously found guilty of a capital or similar serious felony-level offense. One person had spent 27 years in prison for a crime he did not commit; another had been in prison for 23 years.

In addition to the sixteen overturned convictions, another twenty-four people were released from Dallas, Texas, jails and prisons who were wholly exonerated as part of a "fake drug" scandal in 2001. In that scandal, Dallas Police Department narcotics officers provided false statements in court that affirmed positive field testing of materials for narcotics or similar illegal drugs. The actual evidence against these people, however, did not stand up to a stringent application of the falsification principle, fact verification, and the scandal was uncovered. Unfortunately, the scandal was not uncovered until many innocent people had spent time in jail or prison.

Like DNA testing, fingerprinting, i.e., dactyloscopy, is another tool that is presumably verifiable and repeatable that can provide falsification. The correspondence between individuals and their fingerprint patterns was recognized even in ancient times. Fingerprint impressions were used by the Babylonians and Egyptians as personal identification markers. In the nineteenth century, fingerprints were systematically studied and classified. Jean Vucetich developed the first fingerprint classification system for criminal work in 1888 in Argentina. The modern fingerprinting system was then developed by Sir Edward Richard Henry in the United Kingdom in the late nineteenth and early twentieth centuries, and was put to practical use there in 1901. This same system was then applied in the United States in 1903.

Unless a person has a detachable finger that he occasionally loans to others, or a transfer of a fingerprint impression has otherwise occurred, the finding of a person's fingerprint can physically confirm that his corresponding finger contacted the object on which the fingerprint has been found. Consequently, the act of touching need not be seen by eyewitnesses to be confirmed. The fingerprint itself documents the contact. This is, of course, a very powerful investigative tool.

However, fingerprint identification mistakes, as well as DNA test mistakes, due to poor training, poor technique, ambiguous or inadequate procedures, or simple malfeasance, certainly occur. Consequently, the reliability of fingerprint evidence has been called into question.

In 1995, a fingerprint proficiency test was conducted by the Collaborative Testing Service to determine how well fingerprint examiners correctly match fingerprints to the correct individuals. The test was designed by the International Association for Identification, a professional fingerprinting organization. The test was as follows. Four fingerprint cards with all ten fingerprints were given along with seven latent prints. Of the 156 fingerprint examiners who participated in the test, only 44% correctly classified all seven of the latent fingerprints. Concerning the relatively poor outcome of the test, the editor of *Forensic Identification*, David Grieve, was quoted as follows (http://topics.nytimes.com/top/reference/timestopics/people/m/brandon_mayfield/index.html):

> Errors of this magnitude within a discipline singularly admired and respected for its touted absolute certainty as an identification process have produced chilling and mind-numbing realities. Thirty-four participants, an incredible 22% of those involved, substituted presumed but false certainty for truth. By any measure, this represents a profile of practice that is unacceptable and thus demands positive action by the entire community.

A famous example of this problem occurred in 2004 after a terrorist bombing of a train in Madrid, Spain. A number of latent fingerprints were found by the Spanish National Police after the bombing. These latent prints were then forwarded to the U.S. FBI for possible identification.

Using their latest equipment, the Integrated Automated Fingerprint Identification System (IAFIS), the FBI identified a lawyer in Oregon as matching one of the prints. The IAFIS system contains the fingerprints of more than 54 million people. Most of the fingerprint records are associated with people who have some kind of criminal record. However, perhaps 1.5 million of the fingerprints on record are of individuals who do not have criminal records. They are prints that have been collected as a result of applying for a federal job or similar civilian activity.

The FBI initially said that the match was 100% positive. The Spanish Nation Police, however, disagreed with the findings of the IAFIS and concluded that there was not a match. After 2 weeks passed, another person was identified by the Spanish National Police as matching the fingerprint. In May 2004, the FBI admitted they had made a mistake, and the Oregon lawyer was released from jail.

Later in January 2006, a U.S. Justice Department report indicated that mistaken identification by the FBI was due to poor application of methodology by the FBI examiners. As an aside, the Oregon lawyer whose fingerprints were misidentified by the FBI was an American convert to Islam who had married an Egyptian immigrant. The FBI and the Oregon lawyer settled out of court for a considerable monetary sum and the FBI formally apologized in late 2006 for the mistake.

The above event aside, advances have been made in fingerprinting such that minute traces of the grease and oil that make up a person's fingerprint can be analyzed for various organic compounds that indicate something significant about the person's medical condition, lifestyle, or occupation. For example, smokers can be identified by the traces of nicotine metabolites that are left in their fingerprints.

Returning to the subject of falsification, until recently, medical researchers, especially that associated with drugs for the treatment of diseases, tended to publish papers that primarily touted clinical successes. When a drug appeared to cure a disease or significantly diminish its effects, the test results were readily published in professional journals. The good results were then promoted and quoted in business periodicals. This usually led to high profits for the company that underwrote the research, and fame for the principal researchers.

On the other hand, research that revealed negative or undesirable aspects of a drug, such as nonperformance, deleterious side effects, or performance less than that of other currently available drugs offered by a competing company, often was either not reported at all or only reported as required, with minimum details. Negative or deleterious clinical test results, while extremely useful to the medical community and especially to patients, often result in losses or reduced profits for drug companies and little peer recognition for the researchers. Consequently, in the matter of drug testing there has been an economic motivation to overlook the principle of falsification.

The above is an example of how a confirmation bias can develop over time and eventually become institutionalized through repeated positive reinforcement of the behavior.

This was just one of the many shortcomings of published research concerning the possible harmful effects of smoking conducted by a tobacco industry–sponsored research institute. For years this research institute, ostensively staffed with *bona fide* scientific researchers, publicly published reports that failed to find any deleterious effects associated with smoking. It wasn't until their private archives were subpoenaed and made public that it was discovered that while reports discussing the harmful effects of smoking did exist, the research institute simply didn't submit them for publication. There was no requirement, other than good scientific practice, to do so.

In other words, research funded by the tobacco industry that appeared to support the hypothesis that smoking did not cause health problems was published, but research funded by the tobacco industry that appeared to falsify that hypothesis was not published. The subsequent publication record of this research institute confused the public for many years until definitive studies conducted by the Surgeon General's Office concluded that there was, indeed, a direct link between smoking and lung cancer, emphysema, and heart disease.

More recently, a similar reluctance to publish negative data allegedly occurred concerning a drug called Paxil, an antidepressant. In this matter, the State of New York sued GlaxoSmithKline for holding back data that allegedly showed that the drug increased the risk of suicide in teenagers.

A more recent example of confirmation bias involves the drug Zetia, which was initially advertised to reduce both cholesterol and heart disease better than its competition. After Zetia was approved by the FDA, a 2-year clinical trial of 720 patients was undertaken by the manufacturer. Half the population of patients in the trial were given Vytorin, which is a combination of Zetia and Zocor, and half were given Zocor alone. Zocor is also a cholesterol-reducing drug, but is not as good as Zetia. The results of the study were scheduled to be released in March 2007. However, the results were not made public until January 2008, and were not published in a peer-reviewed journal until March 2008. The results indicated that patients taking the Zetia–Zocor combination drug were more or less at the same risk for heart disease as the patients who took Zocor alone.

What occurred is that while Zetia did reduce cholesterol better than Zocor alone, it did not prevent the buildup of plaque on the interior of arteries, which can lead to heart attacks. In other words, it had been presumed that a greater reduction in cholesterol would consequently result in a greater reduction of heart attacks. What the trial showed, however, was that while Zetia combined with Zocor was more effective at reducing cholesterol, it was not more effective at reducing the buildup of plaque on artery walls. In fact, the Zetia–Zocor combination drug may have slightly exacerbated plaque buildup since it was noted that the arterial walls were somewhat thicker when the combination drug was taken than when Zocor was taken alone. (See Kastelein, J., et al., "Simvastin with or without Ezetimibe [Zetia] in Familial Hypercholesterolemia," *The New England Journal of Medicine*, 358 (2008): 1431–43.)

For reasons such as these, a group of scientific journal editors announced that beginning in July 2005, all clinical trials have to be registered in a public database (see http://clinicaltrials.gov) before they are conducted in order for their results to be published. This policy is expected to shift the present bias of publishing only favorable results. By publicly registering before clinical trials are begun, it will be more difficult to hide negative research results. Additionally, a new publication has been initiated, *The Journal of Negative Results in Biomedicine*. This peer-reviewed journal will be a forum where negative or neutral findings can be reported.

One last example of the principle of falsification involves the use of a new type of forensic evidence that is coming into its own: pollen identification. Pollen, of course, is just about everywhere where there are plants. Pollen seems to land and stick on just about every surface. Importantly, some types of pollen are specific to regions, areas, and even individual valleys. In conjunction with plant DNA analysis, some types of pollen can even be traced to a specific plant.

On humans, pollen can be found not only on and inside clothing, but also on a person's head and body hair, and in his nasal and air passages. The longer the time a person spends in an area, the more pollen that is peculiar to that area he will accumulate on his person. Consequently, the types and amounts of pollen found on a person can indicate where he has been, when he was there, and relatively how long he was there.

In 1995, more than seven thousand Serbian Muslims, including women and children, were summarily executed at Srebrenica in Bosnia. Their bodies were initially buried in a mass grave. To hide, contaminate, and confuse the forensic evidence, however, the bodies were shrewdly disinterred from the initial mass grave and moved to a new one.

The Bosnian Serb leaders in power at the time were Radovan Karadzic and his military commander, Ratko Mladic. As of the time of this writing, these two people are still at large. Mladic is still at large. It is rumored he is hiding somewhere in the former country of Yugoslavia. Karadzic has been apprehended and is in custody at the UN Detention Unit in Scheveningen. Both have been indicted by the UN war crimes tribunal at The Hague.

By examining the bodies in the mass grave for pollen, specific and unique pollen types were identified that could have come from only the original mass grave location. The verifiable pollen matches clearly falsified the premise that the victims had been murdered and buried where they were eventually found, as the perpetrators wished it to appear.

At the time of the massacre, the territory in which the original mass burial was located was controlled by Bosnian Serb troops, and was a United Nations designated safe zone. The second site was not a United Nations designated safe zone. The documentation of the change in burial location is extremely important. It helps prove that the deaths of these seven thousand people were premeditated genocide perpetrated and approved by the leaders in charge of the Bosnian Serb troops who controlled the area, and were done in defiance of the existing United Nations safe zone designation.

With respect to the falsification principle, if a person accepts only facts that support the particular hypothesis he treasures, he can overwhelmingly convince himself that the world is flat, that the Holocaust never happened, that astronauts never landed on the moon, that sea shells found at the tops of mountains spontaneously grew there, that Bigfoot roams the Rocky Mountains and only poses for photographers who can't figure out how to focus a camera, that the world began on Thursday, October 23, 4004 B.C. at 9 a.m., or that the planets and the sun travel in perfectly circular orbits around the earth. A perfectly logical case can be prepared to support all of the above arguments, and many more, if just a few falsifying facts are excluded from consideration.

Because of this, in debate or at trial, an advocate of a particular argument will attempt to exclude falsifying information by

- Discrediting the information
- Discrediting the bearer of the information
- Making up special rules or circumstances that explain the information

The last item in the above list is sometimes called an *ad hoc* hypothesis. Generally it is argumentative in nature rather than being based upon a particular scientific principle or verifiable fact that logically applies to the situation. One of my favorite *ad hoc* hypotheses that is often used to explain why extrasensory sensory perception does not stand up to laboratory testing is: "There are unbelievers in here creating bad vibes that are making the instruments not indicate properly."

A famously amusing *ad hoc* hypothesis involves the pseudoscience of biorhythms, the belief that people's performance is governed by several periodic functions that begin at birth and occasionally reinforce or destructively interfere with each other. Biorhythm advocates once claimed, and perhaps still do, that they could use biorhythm theory to predict the sex of babies.

Using standard analytic techniques and methods, Professor W. Bainbridge at the University of Washington tested this claim. In a paper published in 1978, "Biorhythms: Evaluating a Pseudo-Science," he found that biorhythm theory accurately predicted the sex of babies about 50% of the time, which, of course, is no different than just randomly guessing. A defender of the theory, however, proffered the following *ad hoc* hypothesis to explain why the theory didn't work in the study: Some of the cases where biorhythm theory incorrectly predicted the sex of babies in Dr. Bainbridge's study were due to the fact that these cases included homosexuals, who have indeterminate sex identities.

When an advocate wants to exclude falsifying information in a more genteel fashion during a live discussion, he will often resort to arguments such as "you only report the negative things when there are so many good things to report," or "didn't you also find a lot of good things about this," or "you always did have it in for this person," to cajole the opponent into voluntarily limiting the introduction of falsifying information. Despite its obvious intent, the gambit is often attempted because sometimes it works.

Iteration: The Evolution of a Hypothesis

An important feature associated with both the old and the modern version's of the scientific method is iteration. The initial hypothesis iteratively evolves as new information is discovered, verified, considered, and compared to the existing body of information.

Consequently, if a person asks an investigator, "What do you think caused this to occur?" during the first part of the investigation, the person may get one answer. If the same question is asked later, the answer may change, and if he asks again when the investigation is finished, the answer may have changed yet again. The final conclusion may significantly differ from the initial hypothesis.

The fact that a hypothesis evolves as an investigation progresses and more evidence is uncovered often frustrates administrators and reporters who have deadlines. They want to receive a simple answer to their query that will not change, and they want to receive it right now. They don't like waiting for laboratory tests or additional confirming investigative work to be done. In many cases, they already think they know what the answer is. Consequently, they feel any more investigate work is simply redundant and wasteful.

Often they set artificial schedules and deadlines for when they think the investigation should be completed and may attempt to manage the investigation as if it were a well-defined construction project. They want to be the first people to know the correct answer and loathe having to issue a correction or an amendment after their first official pronouncement. They also don't like unexpected evidence turning up that upsets everything that has occurred previously. A changing hypothesis makes them uneasy, arouses their suspicion, and sometimes causes them embarrassment.

They may say something like, "I thought you told me the cause of this was X, and now you tell it is Y. What gives? I have already fired [or arrested, banished, tortured, burned at the stake, etc.] the guy who you said did it." Explaining to people who are unfamiliar with how the scientific method works that new information that sometimes requires a hypothesis to be modified can be trying for both parties.

Some administrative and governmental processes, unfortunately, are not amenable to changes or modifications once they have been put into motion. Administrative processes often cannot change easily as new information is discovered during an investigation. Once an administrative process has begun, starting, stopping, or modifying it can be hard and often has deleterious consequences. Also, once an answer is provided, the follow-up correction or modification is often not heard, believed, or acted upon. First impressions can create an administrative momentum that is very hard to redirect.

This is what occurred during the 1996 Summer Olympic Games in Atlanta, Georgia. On Saturday, July 27, 1996, at 1:20 a.m., a bombing occurred in a public area where a dance was being held. (Jack Mack and the Heart Attack were playing.) One person was killed and 111 were injured. Another person was killed en route to the bombing scene.

Just prior to the pipe bomb's detonating, a 911 call had been made by an unknown person warning of the bomb. A security guard in the area noted a suspicious green knapsack in front of a sound technician's booth. He alerted the local police and was clearing the area around the knapsack when the pipe bomb detonated.

Initially, the security guard was hailed as a hero and was even commended by the president in a statement. Clearing the area just before the bomb detonated likely did save lives and mitigate injuries. On July 30, 1996, however, the *Atlanta Journal Constitution* published an extra edition with the headline "FBI Suspects Hero Guard May Have Planted Bomb." Various news organizations then picked up the story.

The identity of the bombing suspect and the reasons for suspecting him were somehow leaked to the press from one of the investigating agencies. The press then followed closely the progress of the investigation. In some cases, the press appeared to react more quickly to investigation events than did the authorities. For example, the press had already staked out the security guard's apartment and the cameras were rolling when the FBI showed up to serve him a search warrant.

The press camped on the security guard's apartment doorsteps day after day in order to get exclusive, perhaps controversial, video footage and interviews. As the FBI searched the security guard's apartment and other locations where he had worked, took hair samples, took voice samples, towed his pickup truck away for examination, took fingerprints, had dogs smell through his apartment, removed some of his belongings in plastic evidence boxes, and otherwise searched exhaustively for any physical evidence that might link him to the bombing, the networks and news agencies publicly documented, photographed, and televised each humiliating event.

Every night for several weeks, accounts of the search for physical evidence to link the security guard with the bombing were shown regularly on national and regional television networks and published in national and regional newspaper articles. The press repeatedly showed pictures and video clips of the suspect looking suspicious and defensive as he attempted to walk through the swarm of cameras and tape recorders thrust at him. The strain of constant attention from the press while he was simultaneously being questioned and his home and property minutely searched clearly showed.

On their own initiative, the press conducted their own investigation. They interviewed friends, neighbors, and acquaintances who had known him in the past and used this information to publicly dissect and psychoanalyze his life. They aired interviews with various psychologists who described the "lone bomber" criminal profile for the benefit of the viewing audience and how this might apply in this situation. One entrepreneurial news organization even rented an apartment that faced the suspect's home apartment in order to keep its cameras on him around the clock. The attention from the news media was relentless.

One reason given on television for the security guard's being suspected was that he allegedly fit a psychological "hero" profile that was apparently being tested by one of the investigating agencies. Heroes in this sense are people, often in low-echelon or low-status positions, who promote or stage emergency events. Because they position themselves convenient to the emergency, these events then allow them to save people or property and be heroes. They do this to fulfill a strong need for praise and recognition that is otherwise outside their reach.

Heroes had previously occurred in other situations where bombs were discovered. In 1984, for example, a Los Angeles policeman placed a bomb on a public bus, and then was a hero when he was the first person to discover it and save the people on the bus.

With respect to the hero profile, one news organization reported the following:

> They don't have a lot of evidence on him [the security guard] yet, but it may come pretty quickly. Still, they're not breaking out the champagne yet.... This does fit a profile.... There are people—cop groupies—who like to penetrate crimes and then aggrandize themselves. Some even provide interviews afterward. I'm not saying he [the security guard] did it, but it is something the F.B.I. is taking very seriously.

The spectacle carried on until October 23, 1996, when a federal judge ordered the FBI to make public within 5 days the affidavits used to obtain search warrants of the security guard's property. In a subsequent memorandum filed with the U.S. District Court a few days after the court order, the government cited seven reasons why the governmental investigating agency had believed the security guard was a suspect:

1. A knapsack that belonged to the security guard, which a witness reportedly had seen, was similar to the one that contained the pipe bomb and was reportedly missing.
2. It was reported by a witness that he heard a loud explosion and observed smoke in the woods near the security guard's property. (This was cited presumably to suggest experimentation prior to the event.)
3. Prior to the event, the security guard was reported by a witness to have told another person that someday he would become famous.
4. He took an extended break prior to the bombing, for 15 to 20 minutes, which reportedly was unusual for the security guard.
5. Prior to the bombing, the security guard resisted reassignment to another area of the park.
6. About 4 to 6 weeks prior to the bombing and during the sound towers' construction, a witness said the security guard inquired whether they could stand up to an explosion.
7. A witness reported that the security guard was seen at a college with a sketch that looked like a bomb.

Does the reader note the common element in all seven items?

Hint 1: With respect to item 5, do you think it would be more fun to guard a place where a live band was performing, or somewhere else at 1 o'clock in the morning where nothing was happening? Is this suspicious behavior or is it reasonable behavior?

Hint 2: Go back to Chapter 1 and review the initial logic used to conclude that the propane furnace caused the fire and note the similarities.

The document listing the seven reasons, however, also stated, "To be sure, it is quite possible that [the security guard] had no involvement in the bombing. There are numerous suspects and leads in the case entirely unrelated to [the security guard] and there is evidence suggesting that [the security guard] did not commit the crime" (http://query.nytimes.com/gst/fullpage.html?res=9802E3D61639F934A15753CIA960958260&sec=&spon=&pagewanted=all).

On October 26, 1996, federal prosecutors indicated the security guard was no longer a suspect.

Evidence discovered later completely exonerated the security guard and implicated another person. This other person was named as a suspect in the Atlanta Olympics bombing by the Department of Justice on February 14, 1998. The new suspect appeared to also be involved in several other bombings, including some in Atlanta and some in Birmingham, Alabama. In some of these other bombings, people were injured and killed.

The new suspect allegedly was connected to a Christian identity group that was avidly antigay, anti-abortion, anti-Semitic, antigovernment, and so on. This connection allegedly provided a clear motive, as opposed to a profile, that linked all the bombings. Note that the hero profile that caused the FBI to suspect the security guard in the first place did not apply to the new suspect.

This time, however, the investigation had physical evidence that linked the new suspect to the various bombings. Steel plates reportedly used in the Atlanta Olympics bombing were traced back to a metal working plant in rural North Carolina. A person who worked at the metal working plant was a friend of the same person implicated in the Birmingham bombings. The plates were used to direct debris from the explosion.

Other physical features common to the pipe bombs were also noted and traced. In one instance in Atlanta where allegedly two bombs were to be detonated at the same time, one of them did not detonate. It was recovered and was very useful to be compared with the remains of the ones that did detonate. In short, this time they were following the evidence without a confirmation bias.

The new suspect was eventually named to the FBI's Ten Most Wanted Fugitives list, and a $1,000,000 reward was offered. After being named on the FBI list, the new suspect apparently spent 5 years in the Appalachian Mountains dodging the authorities. He became a minor folk hero to those sympathetic to antiabortion, antigay, antigovernment, and anti-Semitic causes until he was apprehended on May 31, 2003, in North Carolina. At the time, he was reportedly scavenging for food in a trash bin behind a grocery store.

On April 13, 2005, the suspect, Mr. Eric Rudolf, pled guilty to four bombings:

- The July 1996 bombing at the Summer Olympics—where one woman was killed and over one hundred were injured
- A January 1997 bombing of an Atlanta office building that contained an abortion clinic—where six people were hurt
- A February 1997 bombing of an allegedly lesbian Atlanta nightclub—where five people were hurt
- A January 1998 bombing at an abortion clinic—where a nurse was injured and a policeman was killed

In return for his guilty plea, Mr. Rudolf avoided capital punishment charges. Instead, he was given four consecutive life sentences without parole. After Mr. Rudolf was sentenced, his sister, Deborah Rudolph, reportedly said, "He believes that the Bible is the history of the white race, and that the other races in the Bible, you know, are just—he would call them 'mud people.'" She further reportedly said that the Olympics were a target because they featured "all these people coming from all different countries and cultures and colors and races and religions, all coming together in one place" (http://transcripts.cnn.com/TRANSCRIPTS/0504/11/pzn.01.html).

This case is a good example of what happens when evidence collection is driven by theory, that is, by a confirmation bias, rather than a theory being driven by the evidence.

One of the more bizarre aspects of this affair occurred during the time the new suspect was a fugitive. The new suspect reportedly visited his gay brother and, over dinner, forcefully quoted a conservative radio talk show host to him. Apparently inspired by this, on March 7, 1998, the brother cut off one of his hands with a portable electric saw and videotaped the event in order to "send a message to the FBI and the media." The hand, by the way, was sewn back on successfully.

With respect to the original suspect, the security guard, despite the sincere apology of the investigating organization, his professional reputation was devastated after having been labeled as a terrorist bomber night after night on the television news and in newspapers and news magazines. The security guard sued various parties and settlements were reached. His law enforcement career and reputation were somewhat rehabilitated, especially after Mr. Rudolf was caught and sentenced. In 2006, 10 years after the event, the governor of Georgia honored him for having saved lives by his actions on the day of the bombing. The accused security guard, however, did not live happily every after. He died in 2007 at the age of 44, having suffered for some years from the effects of diabetes, heart problems, and kidney problems.

This is an example why many investigators should not disclose anything to people outside the immediate investigative team while an investigation is in progress. Only when the investigators are confident that there is no more significant information, and all the significant assessment that can be done has been done, should they publicly unveil their conclusions and assessments.

Of course, a critical reader at this point might pose the following questions:

- Why didn't the investigators in the Atlanta Summer Olympics bombing affair work harder at the beginning of the case to falsify the initial premise?
- Why was the principle of falsification applied so late in the case?
- Why did it take intervention by a judge to apparently put the investigation on the right track?

Lessons Learned

At this point, I could discuss more examples like the Oregon lawyer who was misidentified by a fingerprint as one of the Madrid bomber terrorists or like the security guard who was misidentified as the bomber at the Atlanta Summer Olympics. The point is that boneheaded mistakes are made when investigators allow themselves to be caught up in a confirmation bias, the tendency to prove the hypothesis that they already believe, and forget to apply the principle of falsification.

In the Oregon lawyer event, the FBI appeared to allow the fact that the lawyer was Muslim and married to an Egyptian immigrant influence their judgment in assessing the fingerprint. These facts apparently made an initially close match appear even closer.

Likewise with the Atlanta Olympics matter, even assuming that the psychological hero profile was valid, was it logical to presume that one and only one person could fit the hero profile? And if more than one person could fit the profile, would they all be bombers, would just some be bombers, or would just the one they tested be the bomber? Was the matter finally settled by a federal judge because no subordinate in the agency had sufficient standing to question the judgment of the person applying the hero profile?

As the reader may well note, critical thinking is an investigator's best friend to keep him out of trouble.

More about the Fundamental Method

3

> When people thought the Earth was flat, they were wrong. When people thought the Earth was spherical, they were wrong. But if you think that thinking the Earth is spherical is just as wrong as thinking the Earth is flat, then your view is wronger than both of them put together.
>
> **Isaac Asimov, from *The Relativity of Wrong*, 1996**

More Historical Background

The roots of the scientific method go back to ancient Greece, to Aristotle's elucidation of the inductive method. By the way, Aristotle, who lived from 384 to 322 BC, was Alexander the Great's private tutor. Aristotle died a year before Alexander died. It must have been a point of pride that one of Aristotle's students really did put his education to work to conquer the world.

In the inductive method, a principle or proposition is put forward. The principle or proposition is then tested against the available evidence that has been obtained through unbiased observations and fact gathering. In assessing all the observations and facts, an underlying commonality is shown to exist in all the verified data that demonstrates the principle or proposition.

The Greeks, however, generally preferred the deductive method for investigations and did not often use their knowledge of the inductive method to investigate phenomena. Aristotle, himself, is more famous for having developed the standard deductive form of logical argumentation: the syllogism.

The successes of the Greeks in applying the deductive method to geometry, and their reluctance to actively research, experiment, and gather facts in the field, further contributed to this tendency. At that time, many believed that the nature of the universe could eventually be understood through the power of intellect alone. In other words, if a person were truly smart enough, facts and physical evidence were unnecessary. A person sufficiently smart could figure everything out by just thinking clearly and logically.

Roger Bacon, a thirteenth-century English Franciscan monk, is often credited with initiating the modern scientific method by combining scientific experimentation and the inductive method. He believed that scientific knowledge should be obtained by close observation and experimentation. Friar Bacon experimented with gunpowder, lodestones, and optics, to mention a few items, and was dubbed “Doctor Admirabilis” because of his extensive knowledge.

Friar Bacon studied at Oxford. As most students of his time, he initially studied the trivium course of study, which included grammar, logic, and rhetoric. Later, he progressed to the quadrivium, which included geometry, arithmetic, music, and astronomy. Roger Bacon’s studies included the works of Aristotle, whose writings were a major part of his studies. He received a master’s degree in 1241 and stayed on for a time at Oxford as a teacher himself.

While Roger Bacon attended Oxford, the teachings of Aristotle had been banned at the University of Paris because Aristotle had not been a Christian. (It would have been a neat trick if he had been.) In the early 1240s, however, Aristotle’s works were reintroduced into the curriculum at the University of Paris, and Friar Bacon was asked to be a lecturer on Aristotle in the Faculty of Arts. As was the custom at universities then, all of his lectures about Aristotle were given in classical Latin.

While at Oxford, Roger Bacon had not been interested in science. However, during his time at the University of Paris he met Peter Peregrinus, a scientific experimenter. Bacon said of Peregrinus during this time that he “gains knowledge of matters of nature, medicine, and alchemy through experiment, and all that is in the heaven and in the earth beneath” (*The First Scientist* by Brian Clegg).

The friendship with Peter Peregrinus, who later wrote an important treatise on magnets, at the University of Paris spurred Roger Bacon’s interest in mathematics and natural philosophy. When Roger Bacon returned to Oxford in 1247, it became his life’s work (www.crystalinks.com/bacon.html). One of his important contributions to science in that period was the application of geometry to optics. From this work, he concluded that applying mathematics to real-world problems was necessary to not only quantify what was occurring, but also use it as a predictive tool. Bacon stated: “Mathematics is the door and the key to the sciences.”

Bacon also used lenses to systematically enhance his ability to observe, that is, to extend his personal senses (www.mysteriousbritain.co.uk/occult/roger-bacon.html). This was daring for the time because many people regarded information gained through the senses as unreliable and subject to the machinations of the devil. Most importantly, Roger Bacon planned, executed, and interpreted experiments using the scientific method, much as it is practiced today.

In 1250, Bacon published a letter, the modern equivalent of a paper or scientific article, entitled *De Mirabile Poteste Artis et Naturae* (Concerning the Extraordinary Power of Technical Skill and the Natural Order of Things). In this letter, Bacon reviewed some of his achievements in optics and ideas he had for some mechanical devices. Being a friar, Bacon was a devout Christian. He sincerely believed that scientific investigations would help in understanding the world and, in doing so, would aid in understanding what God had created.

His superiors, however, did not share this view. For his efforts to put knowledge on a verifiable basis, the curious friar was accused of necromancy, heresy, and black magic by the chiefs of his order. As a result of these accusations, he was confined to a monastery for 10 years in Paris so that he could be watched.

During this time of confinement he attempted to persuade Pope Clement IV to allow experimental science to be taught at the university. His efforts failed. He was, however, allowed to teach mathematics in that period and worked on reforming the calendar. His calendar reforms were not acted on by the pope at the time. Three hundred years later his reforms were finally adopted with little change. Even after 300 years, Roger Bacon, however, received no personal credit for the work. Some grudges apparently die hard.

Some of Bacon's scientific proposals and accomplishments include the following:

- He proposed writing an encyclopedia of all the science knowledge by a team of experts.
- He attempted to make science a university subject.
- By experiment, he proposed that the maximum altitude of the rainbow was 42°, which was the most accurate value of that time.
- He proposed using lenses to build telescopes.
- He proposed that the earth was a sphere and that a ship could sail around it.
- He estimated the distance to the sun at 130,000,000 miles.

After Pope Clement IV died, Roger Bacon was imprisoned for another 10 years by the next pope, Nicholas III. Nicholas III also specifically forbade the reading of Bacon's papers and books. This was somewhat moot, however, since the church had generally banned his work from publication anyway. Some scholars believe that Roger Bacon was the original inspiration for the Doctor Faustus legend.

Why all the fuss over a friar who wanted to play in the laboratory? It is because the authorities believed experimental science threatened the public perception of the correctness of theories promulgated by the church. At the time of Friar Bacon, the doctrine of apriorism was the accepted basis for inquiry. All findings had to agree with the fundamental tenets of the church, as interpreted by the various church authorities. Anything that appeared to disagree with these tenets was, obviously, the work of the devil to tempt poor mortals and could not be tolerated.

The contributions of Lord Francis Bacon in the late sixteenth and early seventeenth centuries have already been briefly discussed. After Francis Bacon, several other notable scientists further advanced the notion of the scientific method, notably Nicholas Steno, whose application of the scientific method was the foundation of modern geology.

By rigorously applying the scientific method, Steno showed that sea shells found on the tops of mountains and within rock layers well buried and distant from the sea did not spontaneously grow there, as was popularly believed at the time. Based upon the evidence, Steno argued that these were the shells of real mollusks that had been entrapped there in very ancient times by the deposition of sedimentary rock around them. He also demonstrated that

- The various layers of sedimentary rock had been the bottoms of ancient seas, oceans, or lakes
- The upper boundary of sedimentary layers had once been level with the horizon
- In well-folded rock layers, the tops of the sedimentary layers could be distinguished from the bottoms by the relative particle size: big particles at the bottom and smaller particles on top
- Layers of rock are arranged with the youngest layers on top and the oldest on the bottom

The above findings, based upon experimentation and verifiable observations, form the basis of modern geological stratigraphy.

Steno was the first person to demonstrate with scientific rigor that the earth was much older than the 6,000-year figure predicated upon Bishop Ussher's estimate. Bishop Ussher had based his estimate upon the lists of generations contained in various books of the Bible.

Interestingly, later in life Steno converted from Lutheranism to Catholicism, became a priest, and was eventually appointed a bishop. He gave up scientific work after his conversion. Steno died when he was 48, most likely from exhaustion and malnutrition. In the early twentieth century, Pope Pius XI officially indicated that the theology of the Catholic church did not dispute Steno's findings as a scientist. Bishop Steno was beatified by the church in 1988.

Getting back to the inductive method, a possible pitfall of it, as was well pointed out by David Hume in the eighteenth century, is that the number of observations may be too few or too selective for a true generalization or conclusion to be made. A false conclusion may be reached if the observations or facts are representative of a special subset rather than the general set. This was the case in the flat earth hypothesis already discussed. This is why it is necessary to actively seek falsification rather than patiently wait for enough nonselective facts to accumulate for a true generalization to be made.

In the nineteenth century, George Boole, who lived from 1815 to 1864, systematically organized logic into a symbolic algebra. A professor of mathematics at Queens College, Cork, Ireland, Boole published *An Investigation of the Laws of Thought on Which Are Founded the Mathematical Theories of Logic and*

Probabilities in 1854. This publication led the way for the symbolic logic developed later by Bertrand Russell, Alfred North Whitehead, and Clarence Irving Lewis in the first half of the twentieth century. Boolean algebra, which has rules and symbols reminiscent of ordinary algebra, is used to solve problems in logic much as ordinary algebra is used to solve numerical problems. It is the basis for modern switching theory and electronic computer programming.

One of the more colorful characters who advanced failure investigation methodology in the nineteenth century was a relatively unheralded French detective named Francois Vidocq (1775–1857). Vidocq changed criminal investigation work, almost single-handedly, by emphasizing systematic evidence gathering, evidence preservation, and evidence analysis techniques.

In essence, Vidocq worked to enlarge on a practical basis the amount of physical evidence available upon which an inductive hypothesis could be based. As noted previously, the sureness of an inductive hypothesis improves as the database widens and diversifies. This minimizes the chance of developing a false conclusion based upon a special subset.

Previous to Vidocq, most criminal investigative work depended primarily upon eyewitness testimony. Unfortunately, eyewitness testimony was generally not compared to physical evidence to test its veracity. Instead, the weight given to eyewitness testimony was a function of a person's standing and reputation. People of good reputation and high standing were believed; people of poor reputation or low standing were discounted.

While not logically defensible, this investigative method was easy to administer. Simply ask a few people with good reputations what they thought happened, and then act accordingly. Consequently, most trials were a series of argumentative accusations, denials, and alibis. Very little physical evidence was collected, often it was not properly preserved or documented, and even more rarely was it evaluated scientifically.

Beginning as a spy for the Paris police, that is, a snitch, informer, stoolie, or fink, by 1811, Monsieur Vidocq had become the Surete's first appointed chief. The Surete is France's equivalent of the detective bureau of the FBI or England's Scotland Yard. As a youth, Vidocq was a petty criminal and ne'er-do-well. He reportedly killed a man in a fencing duel when he was 14 years old. As a soldier in the Bourbon Regiment, he bragged about having had fifteen duels and was repeatedly in and out of jail for gambling, forgery, and similar offenses. Later in life, however, Vidocq used his firsthand knowledge of skullduggery to accomplish the following:

- Invented the method of making plaster casts of footprints and impressions at the crime scene to permanently preserve them
- Patented unalterable bond paper and some indelible inks
- Founded the first modern detective and credit bureau
- Developed systematic record keeping of criminal records

- Introduced ballistics into police work
- Employed systematic surveillance as part of the investigative process, effectively employing disguises and subterfuges
- Formalized systematic criminal investigations

After he had served in the Surete and then a 5-year stint as a publisher, Vidocq set up the first private detective agency and credit bureau, *Le Bureau Des Renseignmnets*, whose motto was: hatred of rogues, and boundless devotion to the trade.

In this capacity Vidocq inspired many writers. Sir Conan Doyle based Sherlock Holmes upon Vidocq's exploits as a private consulting detective, especially Holmes's technique of linking physical evidence found at the crime scene to the perpetrator. In Victor Hugo's *Les Miserables*, both Jean Val Jean and Inspector Javert are based upon Vidocq. In Edgar Allen Poe's "Murders in the Rue Morgue," which by some is considered to be the first modern detective story, the narrator refers to Vidocq's methods. First published as a short story in *Graham's Magazine* in 1841, "Murders" provides an interesting description of the method. While it is somewhat wordy, it bears repeating and is quoted below.

> The mental features discoursed of as the analytical, are, in themselves, but little susceptible of analysis. We appreciate them only in their effects. We know of them, among other things, that they are always to their possessor, when inordinately possessed, a source of the liveliest enjoyment. As the strong man exults in his physical ability, delighting in such exercises as call his muscles into action, so glories the analyst in that moral activity which disentangles. He derives pleasure from even the most trivial occupations bringing his talent into play. He is fond of enigmas, of conundrums, hieroglyphics; exhibiting in his solutions of each a degree of acumen which appears to the ordinary apprehension praeternatural. His results, brought about by the very soul and essence of method, have, in truth, the whole air of intuition.
>
> The faculty of resolution is possibly much invigorated by mathematical study, and especially by that highest branch of it which, unjustly, and merely on account of its retrograde operations, has been called, as if par excellence, analysis. Yet to calculate is not in itself to analyze. A chess-player, for example, does the one, without effort at the other. It follows that the game of chess, in its effects upon mental character, is greatly misunderstood. I am not now writing a treatise, but simply prefacing a somewhat peculiar narrative by observations very much at random; I will, therefore, take occasion to assert that the higher powers of the reflective intellect are more decidedly and more usefully tasked by the unostentatious game of draughts than by all the elaborate frivolity of chess. In this latter, where the pieces have different and bizarre motions, with various and variable values, what is only complex, is mistaken (a not unusual error) for what is profound. The attention is here called powerfully into play. If it flag for an instant, an oversight is committed, resulting in injury or defeat. The

possible moves being not only manifold, but involute, the chances of such oversights are multiplied; and in nine cases out of ten, it is the more concentrative rather than the more acute player who conquers. In draughts, on the contrary, where the moves are unique and have but little variation, the probabilities of inadvertence are diminished, and the mere attention being left comparatively unemployed, what advantages are obtained by either party are obtained by superior acumen. To be less abstract, let us suppose a game of draughts where the pieces are reduced to four kings, and where, of course, no oversight is to be expected. It is obvious that here the victory can be decided (the players being at all equal) only by some recherché movement, the result of some strong exertion of the intellect. Deprived of ordinary resources, the analyst throws himself into the spirit of his opponent, identifies himself therewith, and not unfrequently sees thus, at a glance, the sole methods (sometimes indeed absurdly simple ones) by which he may seduce into error or hurry into miscalculation.

Whist has long been known for its influence upon what is termed the calculating power; and men of the highest order of intellect have been known to take an apparently unaccountable delight in it, while eschewing chess as frivolous. Beyond doubt there is nothing of a similar nature so greatly tasking the faculty of analysis. The best chess-player in Christendom may be little more than the best player of chess; but proficiency in whist implies a capacity for success in all these more important undertakings where mind struggles with mind. When I say proficiency, I mean that perfection in the game which includes a comprehension of all the sources whence legitimate advantage may be derived. These are not only manifold, but multiform, and lie frequently among recesses of thought altogether inaccessible to the ordinary understanding. To observe attentively is to remember distinctly; and, so far, the concentrative chess-player will do very well at whist; while the rules of Hoyle (themselves based upon the mere mechanism of the game) are sufficiently and generally comprehensible. Thus to have a retentive memory, and proceed by "the book" are points commonly regarded as the sum total of good playing. But it is in matters beyond the limits of mere rule that the skill of the analyst is evinced. He makes, in silence, a host of observations and inferences. So, perhaps, do his companions; and the difference in the extent of the information obtained, lies not so much in the validity of the inference as in the quality of the observation. The necessary knowledge is that of what to observe. Our player confines himself not at all; nor, because the game is the object, does he reject deductions from things external to the game. He examines the countenance of his partners, comparing it carefully with that of each of his opponents. He considers the mode of assorting the cards in each hand; often counting trump by trump, and honor by honor, through the glances bestowed by their holders upon each. He notes every variation of face as the play progresses, gathering a fund of thought from the differences in the expression of certainty, of surprise, of triumph, or chagrin. From the manner of gathering up a trick he judges whether the person taking it, can make another in the suit. He recognizes what is played through feint, by the manner with which it is thrown upon

> the table. A casual or inadvertent word; the accidental dropping or turning of a card, with the accompanying anxiety or carelessness in regard to its concealment; the counting of the tricks, with the order of their arrangement; embarrassment, hesitation, eagerness, or trepidation—all afford, to his apparently intuitive perception, indications of the true state of affairs. The first two or three rounds having been played, he is in full possession of the contents of each hand, and thenceforward puts down his cards with as absolute a precision of purpose as if the rest of the party had turned outward the faces of their own.
>
> The analytical power should not be confounded with simple ingenuity; for while the analyst is necessarily ingenious, the ingenious man is often remarkably incapable of analysis. The constructive or combining power, by which ingenuity is usually manifested, and to which the phrenologists (I believe erroneously) have assigned a separate organ, supposing it a primitive faculty, has been so frequently seen in those whose intellect bordered otherwise upon idiocy, as to have attracted general observation among writers on morals. Between ingenuity and the analytic ability there exists a difference far greater, indeed, than that between the fancy and the imagination, but of a character very strictly analogous. It will be found, in fact, that the ingenious are always fanciful, and the truly imaginative never otherwise than analytic.

An interesting quasi-legacy of Vidocq is the exclusive Vidocq Society. This nonprofit organization meets monthly on the top floor of the Public Ledger Building in Philadelphia, where the members honor Monsieur Vidocq by applying their forensic skills to unsolved death cases. The society indicates that their members are forensic professionals who donate their talents *pro bono* to cases brought to them that meet their criteria.

In the 1930s, Karl Popper discussed in *The Logic of Scientific Discovery* how the physical laws of science have evolved through a succession of falsified hypotheses. He observed that an initial hypothesis is put forward that has some explanatory value. As verifiable experiences widen, this theory is falsified and then replaced with a new one that has a greater ability to explain the known facts. Interestingly, he noted that each new hypothesis that has greater explanatory utility has greater potential for falsification.

Popper also proposed a test of whether a proposition or hypothesis was actually scientific. A proposition or hypothesis is scientific, he proposed, if it is falsifiable. That is, there must be an experiment, observation, or perhaps prediction that could prove the hypothesis to be wrong, depending upon the outcome or observation. If a proposition or hypothesis does not have the ability to be disproved by some empirical measure, it is not scientific according to Popper.

Moral sentiments, according to Popper, like murder is a mortal sin, are not scientific, for example, because they are not subject to empirical falsification. Such assertions are simply philosophical, religious, or personal moral judgments. Likewise, a theist, that is, a person who believes in God or a god, would agree that the existence of God is not falsifiable because God transcends human knowledge and experience. The belief in magic, miracles, God, gods, spirits, that all life is just a dream, solipsism, and the like are also not falsifiable. The belief that God or a god does not exist is also not falsifiable. These beliefs cannot be tested by experiment and are not capable of being falsified, especially to a "true believer" or a confirmed nonbeliever.

The logic of many conspiracy theories is often constructed such that they cannot be falsified. All contrary evidence or testimony is explained away by the conspiracy itself. For example, all contrary testimony or eyewitness statements were coerced by the conspirators, any contrary evidence was manufactured by the conspirators, the "real" witnesses have all been killed by the inner cabal of the conspiracy, the "real" evidence has been suppressed by the conspirators, and so on. Any falsification evidence brought forward is explained away *ad hoc* by the conspiracy itself using the three rules discussed previously:

- Discredit the information
- Discredit the bearer of the information
- Invent special rules or circumstances to explain the information

There is no point in investigating an accident, failure, crime, event, effect, or phenomenon if the working hypothesis being pursued cannot be verified by facts and evidence, and cannot withstand scrutiny by falsification. This begs the question: If the hypothesis cannot be falsified, what is the point of investigating it?

A Comparison of Deductive and Inductive Reasoning

While the inductive method was well discussed and considered by the Greek thinkers, especially Aristotle, the deductive method was preferred because of its perceived mathematical-like certitude. The deductive method assumes that a general principle, or set of principles, is true, and then constrains or limits the general principle by qualifications until a specific situation or case is concluded.

In other words, the deductive method:

- Assumes the cause, and then logically reasons to a specific effect that results from that cause
- Argues from the general to the particular
- Assumes a general set, and then defines or identifies a specific subset by imposing constraints, qualifications, or special circumstances

Deductive arguments are often arranged in the form of a syllogism. Here is an example of a syllogism:

Major premise: All elephants have four legs.
Minor premise: Bertha is the name of an elephant at the local zoo.
Conclusion: Bertha has four legs.

In the above example, as long as the first two premises are true, the conclusion must also be true. The above is an example of affirming the antecedent, wherein the minor premise asserts that the particular instance, in this case "Bertha is the name of an elephant at the local zoo," equates to the antecedent of the major premise, which was "All elephants have four legs."

Another type of syllogism is to deny the consequent, wherein the minor premise asserts that the instance in question does not equate to the conclusion. Here is an example of this type:

Major premise: All skateboarders are fun loving.
Minor premise: Bob is not fun loving.
Conclusion: Bob is not a skateboarder.

Again, given that the major premise and minor premise are both true, the conclusion follows.

Here is an example of a bad syllogism, wherein the conclusion reached is incorrect. This type of bad syllogism is called affirming the conclusion. It happens when the minor premise equates a specific quality to the conclusion.

Major premise: All skateboarders are fun loving.
Minor premise: George is fun loving.
Conclusion: George is a skateboarder.

What is wrong? While it may be true that all skateboarders are fun loving, they are not the only group that is fun loving. While it may be true that George is fun loving, he may belong to one of the other groups that are not skateboarders, but are still fun loving. Thus, while George might be a skateboarder, he might also be a fun-loving person who is not a skateboarder. Thus, the conclusion may be false.

Another variant of the above incorrect syllogism is called denying the antecedent, wherein the minor premise asserts that a specific instance is not an instance in the major premise.

Major premise: All skate boarders are fun loving.
Minor premise: Fred is not a skateboarder.
Conclusion: Fred is not fun loving.

Again, there are many groups besides skateboarders who are fun loving. Fred may belong to one of these other fun-loving groups. So, the fact that he is not a skateboarder does not allow the conclusion to be made that he is not fun loving. For the conclusion to be correct, it would be necessary to find out if Fred belonged to any other fun-loving groups, and there is no information in the syllogism to that end.

Deductive syllogisms are often not described in terms of true or untrue, but in terms of sound and unsound. This is because the initial premises do not have to be true in order for the argument and the resulting conclusion to be logical, that is, sound. A logical argument is not synonymous with the truth. This is an important distinction and should be remembered when reading the section on sophistry that follows.

In other words, if one or both of the initial premises are not factually true, a conclusion can be arrived at correctly, that is, with no error in logic, but the conclusion may not be factually true. For example,

Major premise: All planets move in perfect circles around the earth.
Minor premise: Mars is a planet.
Conclusion: Mars moves in a perfect circle around the earth.

The conclusion is false, of course, but it was arrived at by legitimate reasoning from the major and minor premises. The major premise is factually false, while the minor premise is factually true. As with the flat earth argument, however, the major premise may appear to be true to an uninformed person who does not have all the facts.

The converse can also be true. That is, an unsound syllogism can be put forward whose major and minor premises are factually true and whose conclusion is factually true, but the conclusion was arrived at by unsound reasoning.

For example:

Major premise: Dogs have four legs. (*factually true*)
Minor premise: Dogs in my town must wear a collar. (*factually true*)
Conclusion: Phaedough must be a dog because he has four legs and a collar. (*Phaedough is indeed a dog.*)

The conclusion is unsound because cats, ferrets, deodorized skunks, and other pets that are taken for walks in this town are also required to wear a collar, and cats, skunks, and ferrets also have four legs. Thus, it cannot be automatically concluded that an animal is a dog because it has four legs and a collar. It might be a dog, but it might also be something else, such as a pet aardvark or a wombat. More information and qualifications are needed to unequivocally distinguish dogs from cats, or from ferrets, wombats, and so on.

The following two simple examples of poor deductive reasoning demonstrate how factually true premises can lead to a factually incorrect conclusion. It is left to the reader to determine what went wrong in the logic.

Major premise: Theft is a felony crime. (*This is true, but theft can also be a misdemeanor.*)
Minor premise: Jim has not been convicted of a felony. (*This is true. However, Jim has been convicted of a misdemeanor.*)
Conclusion: Jim is not a thief. (*False, Jim is, indeed, a thief. He was convicted of misdemeanor shoplifting.*)

Major premise: When the machinery is operated too fast, it fails. (*This is true. It burns out bearings due to lubrication failure.*)
Minor premise: The machine failed. (*This is true.*)
Conclusion: The machine failed because it was operated too fast. (*False. The machine wasn't even being operated at the time. The machine had an electrical short that caused it to fail.*)

As an aside, one of my favorite ersatz syllogisms is from the 1934 MGM motion picture *The Thin Man*. It was spoken by a female character who was angry at her boyfriend for two-timing her, and had packed her bags to leave him. It goes like this:

I don't like crooks.
But if I did like 'em, I wouldn't like crooks who are stoolpigeons.
But if I did like crooks who are stoolpigeons, I still wouldn't like you!

Since it is not always possible to have the information to construct a sound deductive argument, a more common type of deductive argument involves the notion of cogency. A cogent argument is one in which, assuming that all the premises are true, the conclusion is probably true, but not necessarily true. A cogent conclusion is, therefore, conditional.

Consider the following example:

Major premise: Pat is a name equally common to both men and women.
Minor premise: A person named Pat has the condition anomalopia, that is, partial color blindness wherein red and green are not distinguished well.
Cogent conclusion: Pat is a man.

Why is the conclusion cogent? Anomalopia is an inherited condition almost always caused by a specific genetic defect on the X chromosome.

Eight percent of the male population has anomalopia. Women are generally protected from anomalopia because they have an extra copy of the gene on their second X chromosome, and that duplicate gene is generally not defective. Recall that the XY combination of chromosomes makes boys, while the XX combination makes girls.

However, while mostly men have the condition, there are, indeed, a few women who also have anomalopia: slightly less than 0.5% of the female population. Thus, the probability that a person named Pat who has anomalopia is a man is slightly greater than 15 in 16, or about 94%. Conversely, the probability that a person named Pat who has anomalopia is a woman is less than 1 in 16, or about 6%.

In other words, given that the two observations are true, the example conclusion is probably correct, but there is a small chance that it is incorrect.

What if the major premise in the previous example were now restated as follows?

> Major premise: Pat is a name common to men, women, and pets. Of those who are named Pat, 60% are women, 30% are men, and 10% are pets.

How does this new major premise affect the conclusion?

Recall that the minor premise indicates a person is involved. Consequently, the category of "pets named Pat" can be eliminated. The remaining two categories, when proportioned, indicate that one-third of the people named Pat are men and two-thirds of the people named Pat are women.

Some elementary probability theory then indicates that this improves the chances that the person named Pat with anomalopia might be a woman, but the conclusion that Pat is probably a man is still correct. Here is how the arithmetic works out.

Consider 1,000 people who are named Pat. According to the new major premise 333 will be men, and 667 will be women. Of the men, 8% will have anomalopia, or 26.64 men. Of the women, slightly less than 0.5% will have anomalopia, or 3.33 women. The ratio of the two groups who both have anomalopia is 26.64 to 3.33, or 8 to 1. Thus, under the conditions described by the new major premise, a person named Pat who has anomalopia has an 8 in 9 chance of being a man, 89%, or a 1 in 9 chance of being a woman, 11%.

As the above example indicates, cogent or conditional deductive arguments are essentially statistical arguments. Such arguments are improved when several lines of inquiry are used to further limit the chances of being wrong. In other words, the probability that the argument is correct is improved as more information is gathered that could falsify some of the remaining possible ways in which the argument could be wrong.

For example, if the fact that Pat's occupation is Marine paratrooper and the fact that Pat has never been observed wearing female apparel by friends and relations were added, the odds that Pat is a man are improved. Each verifiable gender-related independent fact that fails to indicate Pat is female also further improves the certitude of the conclusion that he is male.

Cogent argumentation is the basis for many crime profile–based methods. Unfortunately, the profile method used to identify likely suspects is often abused and misapplied. (Review the discussion in Chapter 2 of the profile that was used to pursue the security guard in the Summer Olympics bombing case.)

Consider the following statistics published by the FBI in *Crime in the United States, 2002.* Of the people convicted of arson in 2002 who were 18 years old or older:

- 80.7% were Caucasian
- 17.6% were African American
- 0.8% were American Indian or Alaskan Native
- 0.9% were Asian or Pacific Islander

At face value, these statistics might suggest that when arson is committed, an investigator might have a high probability of success in finding the culprit if he simply looks for a likely Caucasian arson suspect. Conversely, any Asian or Pacific Islander suspects can be readily eliminated because they hardly ever commit arson.

However, now consider the following additional statistics put side by side with the above crime figures.

Please note and consider carefully that the FBI data did not contain a Hispanic/Latino category, while the U.S. Census does. Consequently, any Hispanics or Latinos arrested for arson might have been put into either the Caucasian category, the African American category, the Asian and Pacific Islander category, or perhaps the American Indian/Alaskan Native category. How and upon what basis this division was done is not quantified.

In reviewing Table 3.1, and considering how the Hispanic and Latino arson arrests might have been split up, it can be seen that the arrest statistics offer no useful statistical basis for determining the probability of who an arsonist might be. It can be argued convincingly that the racial makeup of arsonists is directly proportional to the racial makeup of Americans in general, and there is no readily apparent data to falsify this conclusion.

As previously noted, an important distinction between deductive and inductive reasoning is that in inductive reasoning, the chain of reasoning begins not with the assumed cause, but with the result or effect, which is already an accomplished fact. By collecting observations and bringing to bear proven principles and facts to logically assess the observations, the cause is

Table 3.1 Arson Arrests Compared to General Population by Race

Race Identifier	Arsonists in 2002	American Public in 2000 Census	Ratio of % Arsonists to % in Population
Caucasian	80.7%	75.1%	1.075
African American	17.6%	12.3%	1.431
American Indian/ Alaskan Native	0.8%	0.9%	0.889
Asian and Pacific Islander	0.9%	1.0%	0.900
Hispanic/Latino		12.5%[a]	Unknown

[a] The U.S. Census notes that Hispanic or Latino people may be of any race.

then determined. But some results or effects can be caused several ways. Not all cause–effect relationships uniquely correspond to each other.

Inductive reasoning does not have the same kind of mathematical certitude associated with it as does deductive reasoning. In more general terms, the problem is somewhat like the difference between a mathematical relation and a mathematical function. In a mathematical function, a specific value of x has associated with it a unique value of y. And, in reverse, there is one and only one value of y for each x. Consequently, if a particular x is known, then a specific y is known without a doubt.

For example, if the function is

$$y = 2x + 6$$

then if x is 5, y must be 16; it can't be anything else. Likewise, if x is 100, then y must be 206, and so on.

In a mathematical relation, however, a particular value of x may have two or more values of y associated with it. Consequently, additional information must be supplied to determine which of the y values are extraneous and which are the solutions.

For example, if the relation is

$$y = [16 - x^2]^{1/2}$$

then if x is 2, y could be either +3.464 or −3.464. Likewise, if x is −3, y could be either +2.646 or −2.646. Additional information is needed to determine whether the positive or the negative number is the solution being sought. Just knowing x in this case does not lock in the value for y. There is no unique one-to-one correspondence that works both forward and backward.

Inductive reasoning does not use syllogisms. It uses a series of observations combined with previous experience and logic to reach a conclusion. The following is an example:

Event: Thomas came to work late this morning.
Observation: The weather and traffic were normal this morning.
Observation: Thomas drives a car that is in good condition.
Observation: Thomas's clothes were wrinkled and his hair was disheveled.
Previous experience: Thomas is known to be prompt and is concerned about having a professional appearance.
Conclusion: Thomas overslept this morning.

Like cogent reasoning, there could be other reasons, perhaps less probable, why Thomas was late. Thus, the initial conclusion can be further strengthened by looking for additional evidence that supports the premise that Thomas overslept, and by also trying to disprove that he overslept, for example, by checking his car to verify he was not in a fender bender, by checking his work schedule to ensure he was not traveling all night and had to sleep at the airport, by checking toll booths to see what time he passed that point on his route, and so on.

Consider the following inductive conclusion and the evidence upon which it is based:

Conclusion: The 1-gallon buckets of latex paint supplied by Acme do not cover at least 800 square feet as advertised.
Observation A: Of the 50 buckets of Acme latex paint randomly selected from a lot of 1,012 buckets and tested by an independent laboratory under the conditions specified on the label, the average coverage was 730 square feet per bucket with a standard deviation of ±60 square feet.
Observation B: Ten professional painters used Acme latex paint and found that it did not cover the 800 square feet per bucket as advertised. Their estimates of coverage were about 10% short.
Observation C: Five guys I know used Acme latex paint and at least four of them said they needed more paint than they thought they would when they painted the exterior of their houses.
Observation D: A salesman for a competitor of Acme latex paint told me that Acme exaggerates how much area one bucket of paint will cover, but that his company's paint actually covers 5% more than is listed on the bucket.

All the above observations support the conclusion that a 1-gallon bucket of Acme latex paint does not cover 800 square feet as advertised. However, some observations support the conclusion better than others.

How much support is needed to justify a conclusion is based upon two factors: the probability that the conclusion is correct and the consequences assumed if the conclusion is not correct as compared with the consequences assumed if the conclusion is correct. In other words, the amount of evidence needed for a conclusion, or decision, depends upon the risk and consequences of that conclusion or decision being wrong.

If a person is simply trying to economically paint his garage, either observation C or D may be sufficient for him to decide whether to purchase Acme or the competing brand. If his conclusion turns out to be wrong, the risk is simply that some extra paint will have to be bought.

On the other hand, if a person is suing Acme for false advertising, both A and B may be needed as well as a second report from another independent laboratory that sampled a different lot of paint.

In short, when using inductive reasoning, the type and degree of certitude of the evidence must be commensurate with the importance of the conclusion. Perhaps, it is somewhat like making a part in a machine shop. How close do the tolerances have to be? As close as they need to be to fit well, operate correctly, and not cost too much.

Apriorism and Aposteriorism

Apriorism is derived from the Latin prepositional phrase *a priori*, which means "from before" or "from the previous." It is the belief that the underlying causes for observed effects are already known, or at least can be deduced from fundamental principles. Under apriorism, especially during Bacon's time, if a person's observations conflict with the accepted theory, then somehow the person's observations must be imperfect. They may even be the manifestation of the devil tempting a foolish mortal.

For example, in Friar Bacon's time it was assumed *a priori* that the sun revolved about the earth and that all celestial motions followed perfect circles. Thus, all other theories concerning the universe and the planets had to encompass these *a priori* assumptions.

This led, of course, to many inaccuracies and difficulties in accounting for the motions of the planets. Why then were these things assumed to be correct in the first place? It was because they were deemed reasonable and compatible with accepted religious dogmas. In fact, various Bible verses were cited to "prove" the validity of these assumptions.

Because holy scriptures had been invoked to prove that the earth was the center of the universe, it was then reasoned that to cast doubt upon the assumption of the earth being at the center of the universe was to cast doubt by inference upon the church itself. Because the holy scriptures and the church were deemed above reproach, the problem was considered to lie within the person who put forward such heretical ideas. In other words, if the data did not match the approved or accepted theory, the data were wrong.

Fortunately, we no longer literally burn people at the stake for suggesting that the sun is the center of our solar system or that planets have elliptical orbits. The modern scientific method does not accept the *a priori* method of inquiry. The modern scientific method allows evidence to drive theories, rather than allowing theories to drive evidence.

Apriorism in disguise, however, is certainly a problem in failure analysis. Apriorism can occur when an investigator is committed to a particular theory of how the failure occurred or perhaps who committed the crime, and then sets out to prove his hypothesis. In other words, he already has decided the answer; he is merely obtaining sufficient evidence to comply with form, or regulations.

This is the germ of the classic "hunch" plot device used in some detective novels. The protagonist, the intrepid detective, spends most of the book trying to prove that his initial hunch concerning the crime is correct and that the hunch formulated by his antagonist, usually an insipid police captain who won't listen or a corrupt or politically ambitious district attorney, is wrong.

Consequently, the investigator may not investigate with an open mind and does not allow the evidence to lead him wherever it may. Evidence that may prove there was a cause different from what the investigator initially has proposed is overlooked, or not looked for at all. Evidence that is found that suggests a different cause is interpreted with special rules, usually tacitly dependent upon unproven but plausible assumptions, to make the evidence fit the preconceived cause.

A similar effect also occurs when the supervisor of an investigator has already made up his mind what caused the failure or who committed the crime, as in the detective novels. When a progress report is made about the investigation and the supervisor disagrees with the direction the evidence is taking the investigating team, the supervisor may direct that certain lines of inquiry that appeared otherwise fruitful be stopped or limited, and that other lines of inquiry that are not being driven by the evidence be emphasized.

Perhaps the investigation is unearthing embarrassing information, perhaps the supervisor has a political agenda and needs this investigation to reach a certain conclusion, or perhaps the supervisor has already made administrative decisions predicated on his personal theory being correct. In any case, a version of apriorism occurs when the work of otherwise competent investigators is modified and redirected to conform to the expectations of superior authority. At that point, the investigation is said to be managed to achieve the desired conclusion.

In this regard, there is a tale from ancient Greece that comes to mind called "Procrustes' Bed." Procrustes was an innkeeper and a serial killer, which is an efficient combination. Procrustes would wait alongside the highway and entreat travelers to stay at his inn by telling them how comfortable the bed was and how the bed would perfectly fit the tired and footsore traveler.

Hidden at the head of Procrustes' bed was an ax, and hidden at the foot was a rack. When an unsuspecting traveler lay down and tried the bed,

Procrustes would seize and rob him. If the traveler were tall, Procrustes would trim off his head, feet, and ankles until he was just the right size to fit the bed. If the traveler were too short, he lengthened him on the rack until he was the right size. In this way, everyone could fit into Procrustes' bed. The tale says that this went on until the hero and mythical founding king of Athens, Theseus, tricked Procrustes into lying on his own bed and then turned the tables on him.

Apriorism is like Procrustes' bed. All evidence, no matter how tall or short, can be fitted comfortably into a predetermined bed if it is tortured enough.

One of the thornier problems for an outside investigator or consultant in the reconstruction of a failure or catastrophic event is the insidious application of *a priori* methodology by a client. This occurs when the client hires an outside expert to find out only information supportive of his particular position or theory.

The client will not specifically state what findings are to be made by the expert, but he may direct the investigator toward items and evidence that he knows supports the conclusion he wants, and misdirect the investigator away from evidence that might support a finding the client doesn't want. The client might say something like, "No need to go through that stuff. We already looked and there is nothing there," or "We already know that that doesn't apply."

While the argument may seem innocent enough, and may give the appearance that the client is just trying to be efficient with the consultant's time, the ploy can also be used to bias the investigator's conclusion because only data that support the desired conclusion are being examined and all other data are being ignored. As with the previous graph analogy, if enough data points are erased or ignored, almost any curve can be fitted to the remaining data points.

A second version of this same *a priori* problem occurs when the client does not provide all the observational data to the investigator for evaluation. Important facts are held back, or the consultant is not given enough time or information to properly do his job. This similarly reduces the observational database and enlarges the number of plausible hypotheses, which might explain the facts.

A third variant of the same *a priori* reasoning is when the investigator becomes an advocate for his client. In such cases, the investigator assumes that his client's position is true even before he has evaluated the data. This occurs because of friendship, sympathy, or a desire to please the client in hopes of future assignments. There is nothing like giving a paying client exactly what he wants to make him happy and get more business.

To partially guard against this, most states require that when employed as a private consultant, licensed professional engineers can accept payment only on a time and materials basis in accordance with published rates. They may not accept bonuses or contingency fees based upon how well their investigative findings please the client. This is essentially selling the engineer's seal. Similar conflict of interest laws and regulations sometimes apply to other state licensed professionals. While some professional groups or organizations that are not state licensed usually advocate similar ethical standards, they often do not have the legal authority, or the will, to enforce them.

If court or legal proceedings are involved, it is common for both of the adversarial parties to question and carefully examine experts to learn how they reached their conclusions. This is done during both pretrial activities, such as in deposition, and the trial itself in accordance with either the Federal Rules of Evidence or the particular state's rules of evidence.

During court examination by the attorneys for each party, the judge or jury can decide for themselves whether the expert is biased. During such court examinations, the terms of hire of the expert are questioned, his qualifications are examined, any unusual relationships with the client are discussed, all observations and facts he considered in reaching his conclusions are questioned, including his reasoning and analysis, and so on.

In the legal system, while an expert is considered a special type of witness due to his training and experience, he is not held exempt from adversarial challenges. He cannot simply spout a conclusion and presume it will be accepted at face value. It is rare for an expert's conclusion to go unchallenged in court.

To form a conclusion, the expert can examine any evidence, including some evidence that would not be admissible in court under the applicable rules of evidence. However, if the expert is using inadmissible evidence to form his opinions, he cannot divulge the inadmissible evidence to the jury or the court while explaining his opinion. An expert cannot be used in this way to get otherwise inadmissible evidence before the jury through a "back door."

While not perfect, this system does provide a way to check the biases and *a priori* assumptions an expert may have.

Opposing apriorism is aposteriorism. Aposteriorism comes from the Latin prepositional phrase *a posteriori*, which means "from afterwards" or "from what comes after." It is the belief that the underlying cause for an event or condition is determined by critical observations of its effects, that is, by examining the facts and circumstances associated with the event or condition. Aposteriorism admits that a particular event could be caused by a variety of antecedents and that only when the facts and circumstance are examined can some potential causes be eliminated and some affirmed.

Consider the following. There are many ways that a house can catch fire and burn down. In order to distinguish arson from an accidental cause, such as an electrical short, it is necessary to at least examine the remains of the house, determine the point of origin of the fire, interview the owner, and examine the circumstances associated with the fire. By doing this, some ways in which the house may have caught fire can be eliminated, and some can be affirmed.

For example, if shorting occurred to cause the fire, wiring that exhibits primary shorting should be present at the point of origin in places consistent with shorting, for example, old wiring or substandard connections. Arson indicators should be absent from the point of origin.

On the other hand, if an arson were committed, the telltale signs of arson occurring first should be present, such as the remains of an accelerant at the point of origin, ignition energy being mysteriously available where none was known to exist, a timer present at the point of origin, and so on. Accidental fire ignition indicators should be absent or secondary around the point of origin.

If enough evidence is available and gathered, and the analysis of the evidence is sound, the way in which the fire actually did occur can be separated with certainty from the ways in which the fire could have occurred but didn't. The only way to find this evidence, however, is to diligently look for it. In this regard, there is no substitute for fieldwork, that is, evidence gathering.

One of the telltale signs of implicit apriorism is a small amount of fieldwork, coupled with a large amount of armchair theorizing and analysis. Likewise, one of the telltale signs of aposteriorism is a lot of fieldwork, coupled with a small amount of analysis. *Very little analysis is needed when a chorus of converging facts speak for themselves.*

With respect to initiating an investigation and avoiding *a priori* biases, perhaps the best advice is that given by the character Siddhartha to his friend Govida in the book *Siddhartha* by Hermann Hesse.

> When someone is seeking, said Siddhartha, it happens quite easily that he only sees the thing that he is seeking; that he is unable to find anything, unable to absorb anything, because he is only thinking of the thing he is seeking, because he has a goal; but finding means: to be free, to be receptive, to have no goal. You, o worthy one, are perhaps indeed a seeker, for in striving towards your goal, you do not see many things that are under your nose. (*Siddhartha*, Hermann Hesse, MJF Books, 1951)

Sophistry

In ancient Athens in the fifth century B.C., sophists were teachers of philosophy who provided instruction about debate and rhetoric. Rhetoric is the art of speaking effectively and persuasively. It is a combination of argumentation tactics, logic, presentation, enunciation, cadence, delivery style, and vocabulary. Due to the importance of public speaking then, especially with its application to Athenian politics and law, rhetoric was an important part of Greek education.

Sophistry in the modern sense of the word, however, is argumentation whose point is to win over an audience by any means. This can be done by appealing to the audience's emotions, their traditions, their local sense of what is plausible, any religious or philosophical beliefs held by the audience, or any underlying prejudices or biases that they hold dear.

Sophistry does not require the use of logic, proven scientific methods and principles, or verifiable facts. However, some verifiable facts or scientific principles are often interwoven into a sophist argument to make it appear credible and scientific. In well-crafted sophistry, especially during the give and take of debate, an illogical or irrational assumption buried unobtrusively in the sophist's argument may be difficult to spot quickly. This may allow the sophist enough time to win over the audience and win the argument before the other side can detect the logical flaw.

Because of this, some dictionaries define sophistry in the modern sense as a deliberately invalid argument that is ingeniously put together to persuade or convince an audience of some issue.

Sophistry is effective when the "truth" of an argument is decided by an audience at a specific time. Sophistry is ineffective when it is subject to expert peer review; scientific standards, tests, and experimentation; or logical scrutiny applied over an indeterminate period of time. If the audience agrees with the argument when the time for decision making occurs, then the argument becomes true.

Consequently, sophists point out that the "real" validity of an argument or the "actual" truthfulness of an argument is unimportant. What is important is winning over the audience. Sophists believe that the audience defines what the truth is. In essence, sophists believe in a relative truth or perceived truth rather than an absolute truth. The truth of a sophist's argument does not have to be consistent with the physical laws of nature or verified facts. It just has to be believed by the audience.

So what does sophistry have to do with failure investigations?

Consider the fact that juries decide whether a particular failure was an accident, a deliberate act, or a crime. Juries decide who is at fault or not at fault in a lawsuit. Juries decide what is good medical practice or quackery. Juries decide who is guilty or innocent in a criminal trial. To an extreme, if an attorney can convince a jury of twelve men and women that the moon is made of green cheese, then legally the moon is made of green cheese: case closed.

Likewise, when industrial failures occur, the management of a company ultimately decides whether a certain root cause finding is acceptable and how it will be corrected. Like juries, managers and management groups have emotions, prejudices, and biases to which a consultant or investigative group can appeal, or perhaps pander. Consequently, the findings of some investigations are much easier for management to accept than others.

Some company management teams, for example, are much more likely to accept the findings of an investigation if they can easily afford the recommended corrective actions, and no one in the management team is found to have either directly or indirectly caused the failure.

On the other hand, investigators who have to present a report that discusses a failure that was the direct result of a management decision or blunder, especially by a specific manager who sits in judgment of the final report, may have a harder time of it. Further, it becomes even harder for the investigator when the necessary corrective action is expensive, is embarrassing, or has undesirable legal consequences.

With respect to crimes, given exactly the same evidence, arguments, and circumstances, a jury chosen from city dwellers might acquit a defendant while another jury composed of farmers might convict, and vice versa. Thus, if the psychological profile of a jury panel is understood, a more convincing argument can be crafted to fit the evidence and the jury's predilections.

This is why jury pool psychologists have become important in high-profile cases. Given sufficient data concerning the background of a specific jury panel and the arguments and counterarguments that will be proffered, a jury pool psychologist can assess the probability that a case will be won even before the trial takes place.

For example, suppose there is a trial and the physical evidence clearly indicates that Mr. Smith is a thief. His fingerprints were found on the merchandise, he was seen by several witnesses carrying away the goods, the stolen goods were found in the trunk of his car, he has no explanation of why he had stolen goods in his car, he has no alibi for the time when the crime was committed, and he confessed to the crime when initially apprehended. If Mr. Smith's advocate, however, convinces the jury, by any means, that Mr. Smith is innocent and the jury acquits him, Mr. Smith is indeed innocent in the eyes of the law.

Likewise, suppose the prosecution is able to convince the jury that a Mr. Jones committed a crime, despite the fact that there is no scientific evidence linking him to the crime, that there are logical inconsistencies and misinterpretations of the physical evidence, and that the defendant has a reasonable alibi. If the jury convicts Mr. Jones on circumstantial evidence that could have equally applied to dozens of other people, he indeed is guilty in the eyes of the law despite the lack of scientific and logical rigor.

In that case, Mr. Jones's only recourse is to appeal the jury's decision to a higher court. If he receives a hearing in the higher court, he will have to trust that the higher court will not be as easily swayed by emotional arguments and will give greater credence to sound logic. Unfortunately, appeals are usually based upon procedural issues rather than logical or scientific ones.

As noted previously in this chapter, this is exactly what occurred to a number of people who were scheduled to be executed in Illinois. They were convicted unanimously by a jury of their peers, and their appeals had been duly considered and denied. Fortunately, their lives were spared by the convincing nature of DNA identification evidence.

One of the argumentation techniques used in sophistry is an unclear definition of the subject matter, i.e., using unclear descriptive vocabulary when talking about a particular subject. Taking advantage of this, an advocate can address two different aspects of a topic at the same time. The audience members can then pick and chose what they wish to hear depending upon each person's personal prejudices.

Another technique is to play upon semantics, that is, to apply shades of meaning not normally intended for a word or phrase, and then to make it mean something useful for your argument and detrimental to your opponent's. Here is an example:

Many states have loans and scholarships for poor students.
Jim's grade point average is barely above a D.
Fred's grade point average is an A+.
Jim is better qualified than Fred for a loan or scholarship because he is a poor student.

Here is another example of sophistry. Suppose an attorney, Mr. Charles, makes many specious legal motions during a trial that he knows will be denied, but he makes them anyway. Over and over he tries the court's patience as he objects to evidence or proceedings that the judge denies. Finally, on a particular point of law that is extremely important to his case, Mr. Charles makes one more objection.

In a sidebar, Mr. Charles tells the judge that the judge is obviously prejudiced against him since the court record will clearly show that the judge has denied 100% of all his motions, while the judge has ruled in favor of his opponent 100% of the time. Mr. Charles will then feign exasperation, accuse the judge of bias against him and his client, and plead for a scrap of fairness from the judge. To prove it, the judge should certainly grant Mr. Charles at least this one motion.

The question is then: Will the judge recall on the spot that all the motions put forward by the other attorney were well grounded and deserved being granted while the motions put forward by Mr. Charles were not? Will the judge be swayed by the apparent percentage figures cited to "prove" that the judge has been unfair? Will the judge promptly recognize the false assumption underlying the argument by Mr. Charles that "fairness" means having an equal number of motions granted to both sides?

Will the judge further think about how this fairness issue will look to members of the press, who are watching in the back of the courtroom and seem to be pulling for Mr. Charles? Will the judge interpret the looks of consternation on the faces of the jurors to mean that he does appear to them to have been unfair?

Or, will the judge consider only the legal merits of the issue under discussion at the moment? If his thinking about the issue, pro and con, was evenhanded before the sidebar and his observation of the jury and the press, will any of these other factors be the slight nudge that pushes his decision one way or the other?

This is one of the insidious facets of sophistry that make it a useful tool for an advocate. When the evidence is nearly equal on either side of an argument, when a decision has to be made, and when there is pressure to promptly make the decision, often the underlying deciding factor is an unspoken, illogical, emotional one. It may be a factor that makes the decider emotionally comfortable with the decision.

The following are some general characteristics of sophistry:

- Ambiguity and equivocation. This occurs when the usual meanings of words and terms are disregarded and unusual definitions and tortured meanings are substituted. Example: *Amputation* and *divorce* are both terms that describe separating and dividing, but a person cannot seek compensation for a divorce through insurance plans that pay for amputation of limbs.
- Nullification. This is where a line of reasoning is attacked by asserting something else is true that cannot be proven or is simply argumentative. Example: "Mr. Jones, is it possible you could be wrong?" or "Isn't it true, Mr. Jones, that if this were all a dream, your argument would mean nothing?" or "Could it be true, Mr. Jones, that the devil is making you say these bad things about my client?" or "Ten years ago one of your laboratory tech employees was convicted of a felony, so any testing from your laboratory cannot be trusted."
- Substitution of arbitrary rules for judgment and context. This occurs when the sophist demands that certain rules that he puts forward be held absolute and inviolate despite the circumstances or conditions. In the previously noted example involving the frivolous motions of an attorney, Mr. Charles, the attorney put forward tacitly a "rule" that fairness means that the same percentage of motions granted to the prosecution should be granted to the defense. This fairness rule invented by Mr. Charles completely disregards whether any of the motions were lawful.
- *A priori* criteria for evidence. This is where the sophist requires that all evidence that will be presented or discovered in the future must be weighed and evaluated according to rules that the sophist decides should apply. Consequently, any evidence that does not fit the rules the sophist has established should not be considered. I trust that the reader will recall the story of Procrustes' bed.

- Shifting of the burden of proof. This occurs when the sophist demands that every hypothesis the sophist proposes be disproved by the other side, and if the other side cannot do so convincingly, then the hypothesis is automatically true. In other words, "If you can't prove me wrong, I must be right." Or when the sophist attempts to transfer the burden of proof of an argument that is normally presumed by one side, to the other side. For example, a person is usually presumed to be innocent until proven guilty. A sophist might turn the argument so that the person is presumed guilty until proven innocent.
- Requiring unreasonable amounts of ironclad proof. This is where the burden of proof requirement is so extreme, no amount of evidence of any kind can prove the issue to the satisfaction of the sophist. Consider the following exaggerated line of questioning involving an armed bank robbery:

  Advocate: Did my client have a gun?

  Witness: Yes.

  Advocate: Do you know what kind?

  Witness: No, not specifically. It might have been a 0.44 Colt automatic.

  Advocate: Was it loaded?

  Witness: I don't know for sure.

  Advocate: Are you familiar with guns?

  Witness: Yes. In the military I earned a sharpshooter's medal for automatic handguns.

  Advocate: Are you a certified handgun identification expert?

  Witness: No, I am not such an expert.

  Advocate: Could it have been a metal toy gun that looked authentic?

  Witness: I suppose it could have been a very good replica.

  Advocate: Did you try to determine if it was a replica?

  Witness: No. He pointed it at me and told me he'd blow my brains out if I moved.

  Advocate: Did you ask him if it was a replica gun?

  Witness: No. I didn't think it was a good idea at the time to engage in conversation concerning his potential collection of replica guns.

  Advocate: Did you think he meant it when he said he would blow your brains out?

  Witness: Yes.

  Advocate: Are you a licensed psychologist and were you able to render an on-the-spot professional opinion that he meant it without a previous psychological interview?

Witness: No, I'm a bank teller, not a shrink.

Advocate: So, since you are not a certified gun identification expert, and since you didn't ask him if it was a replica, and since you are not a licensed psychologist and could not professionally assess his state of mind as to whether he would really kill you or was just joking with a replica gun, what other proof do you have that my client was allegedly threatening you?

Witness: None.

A similar argument might be used in a liability case. It would go like this:

Advocate: Do you agree that a person's life is infinitely valuable?

Witness: Of course.

Advocate: Are you aware that people are sometimes killed in the cars your company manufactures?

Witness: Yes.

Advocate: Don't you think it would be a good thing to spend as much money as it takes on safety until no one is ever killed in any of the cars you manufacture?

Witness: We certainly spend a lot of money on safety research and are doing our best to make them as safe as possible.

Advocate: But people are still being killed in your cars. Your company obviously is not doing a good job and is not spending enough.

- Judgment by secret criteria. This occurs when the sophist sets himself up as a judge or self-appointed critic, and then states something like, "Do you expect rational people like me to actually believe you?" The inference is that the argument is illogical and that the sophist is an objective judge as to whether the evidence is believable.

The Method of Exhaustion

The method of exhaustion, sometimes called the process of elimination, attempts to determine the cause of a failure or event by proving what did not cause the failure or event. When all the possible ways in which the event or failure could have occurred are eliminated until there is only one possibility left, the remaining possibility is presumed to be the cause.

Logically, it works something like this. Suppose event Z can only be caused by one of six possible causes acting alone: causes A, B, C, D, E, or F.

Suppose then that there is evidence that verifies that causes B, C, D, E, and F could not have been involved. Then without having to find evidence to support cause A, cause A is presumed to have caused event Z by the process of elimination. In sum, since there are only six possibilities, and because five of them are eliminated by the evidence, the single remaining cause must be the one that caused event Z, even though no direct evidence was found to support it.

The method is summed up nicely in a quotation by Sir Arthur Conan Doyle's character Sherlock Holmes:

> When you have ruled out the impossible, whatever is left, no matter how improbable, must be true.

In theory, the above sounds like a good plan, especially when it is easier to disprove, that is, falsify, all the alternative hypotheses than it is to find evidence to support (along with evidence that fails to falsify) the remaining hypothesis. The method of exhaustion, however, only works well when all of the possible modes of failure have actually been accounted for in an exhaustive list. If the list is not truly exhaustive, the method is significantly flawed.

In the previous example it was assumed that event Z could be caused only by one of six causes acting alone: A, B, C, D, E, or F. However, what if one potential cause was not included in the list? In that case, if causes B, C, D, E, and F were all eliminated, this would still leave two possible causes: cause A and the unknown cause. On the face of it, this means that this application of the process of elimination might only produce an answer that has a 50% chance of being correct.

Further, while causes B, C, D, E, and F may have all been eliminated by the evidence that they acted individually to cause event Z, what if causes B and E can act in concert to cause effect Z, or perhaps C, D, and F all acting in concert? The evidence only eliminated B, C, D, E, and F acting individually.

In the Michelson–Morley experiment, which was discussed in a previous section, aether must exist, scientists of the time argued, because there was just no other way to account for the transmission of light through empty space. As it turned out, however, the assumption was false. There was indeed another way to account for the transmission of light through empty space: a previously unknown phenomenon involving the propagation of electromagnetic waves. Speaking rhetorically, the scientists who argued for aether on the basis of "it can't be anything else" did not know what they did not know. They presumed that light must act only in ways that they already understood.

Consequently, the process of elimination sometimes suffers from the fact that when the list of all possible failure modes is prepared, an item may be inadvertently left off the list. Perhaps no one knows about it, or assumptions tacitly made in the preparation of the list inadvertently overlooked important sets of possible failure modes.

In arguments where sophistry is used to sway an audience, an advocate may deliberately omit a potential failure mode from the supposed exhaustive list of failure modes, hoping that no one will notice the omission. The advocate will then eliminate by *bona fide* evidence all the items on the list but one, the one he wants to be the cause. Without furnishing any further evidence to support or falsify the one remaining item, the advocate will then conclude by the process of elimination that the remaining item must be the actual cause.

Thus, not only does the advocate deflect scrutiny from the cause he wishes to defend, the one omitted from the list, but he misdirects focus to another cause using an apparently plausible logical argument. This requires the defender to not only furnish the evidence to dismiss the item the advocate concluded was the cause, but also to detect what has been omitted from the list and then find evidence to implicate the omitted item as the actual cause.

This tactic often confuses the audience, who now must weigh and assess three lines of evidence: (1) the process of elimination evidence presented by the advocate, (2) the evidence explaining why the process of elimination logic presented by the advocate is incorrect, and (3) the evidence that supports and verifies the actual cause of the event.

In several of her more famous detective novels, Agatha Christie used the process of elimination logic as a red herring to fool the reader. In one novel, the reader was led to tacitly assume that there was only one murderer, when in fact there were several murderers. The reader was led to assume that if evidence implicated A, the same evidence eliminated B. In another, the murderer actually killed himself early in the story and contrived that others would die after his death. Using the process of elimination, the reader assumed that the murderer would be the person to die last. In a third story, the murderer was the person narrating the story who assisted in the investigation. Again, the reader was led to assume that the real murderer would not honestly assist in providing evidence that would eventually implicate himself.

Because of these pitfalls, the method of exhaustion should be used with great caution and care. Even if there is evidence to conclusively eliminate all but one item on the list, it is good practice to not just accept the logical conclusion. An honest effort to find evidence to support or falsify the remaining item on the list should still be done.

Some investigators include an unknown cause in the list of possibilities and will attempt to find evidence to support or falsify the unknown cause just like any other possibility. If such an approach had been used with the Michelson–Morley experiment, the null results would have indicated right away that a very important but previously unknown phenomenon existed that they hadn't considered.

Coincidence, Correlation, and Causation

Sometimes two events occur closely in time such that the first event is presumed, somehow, to have been the cause of the second event. In other words, because event B happened right after event A occurred, it is presumed that event A somehow caused event B. A cause-and-effect relationship does characteristically have an ordered time sequence. However, the converse of this is not necessarily true. An apparent ordered time sequence, that is, coincidence, is not sufficient by itself to prove that a cause-and-effect relationship exists.

For example, suppose a baseball player has a drink of a particular type of soft drink just before he hits a home run. A week later, he has another drink of the same soft drink and hits another home run. He concludes from these two, separate sequenced events that this is his lucky soft drink. To ensure future success in hitting home runs, he then lays in a few cases of the stuff and drinks it liberally before batting.

Despite appearances, coincidence is fundamentally a random effect that involves independently occurring events. If A occurs, B may or may not occur right after A, and vice versa. Because A and B are independent events, A has no cause-and-effect connection to B. Probability dictates, however, that occasionally when A occurs, B will occur closely afterwards.

People will sometimes argue that coincidental events must be related in a cause-and-effect relationship by noting how improbable the coincidence is. Thus, the argument goes, there must be a cause-effect relationship because the two events occurring in sequence are simply too improbable to occur by mere coincidence.

This argument is great fodder for the previous section on sophistry.

Consider the law of large numbers, however, and how it applies to coincidences. The law of large numbers states that even if something has an extremely low probability of occurrence at any given time, given a sufficiently large number of occurrences in which the event could occur, all things that are possible will eventually occur. The only things that will not occur, given sufficient time and opportunity, are those things that are genuinely impossible.

For example, let's say that the chance of a particular coincidence occurring to a person is one in a million. Since there are about 300 million people in the United States, then on the face of it a one-in-a-million event will happen to 300 people in the United States. Likewise, since there are about 6 billion people in the world, a one-in-a-million coincidence will happen to 6,000 people in the world.

Likewise, the chance of winning the top prize in a state lottery is perhaps 1 in 10,000,000. The odds are low for a specific person to win the lottery, of course, but this doesn't mean that winning doesn't happen. In fact, someone wins the lottery nearly every week.

Consequently, the fact that an event, or sequence of events, has a low probability of occurrence is not *prime facia* evidence that it is not a coincidence. Low-probability events occur unnoticed around us all the time because thousands, if not millions, of events happen around us every day. Further, many supposedly low-probability coincidences even occur in clusters. Some people, for example, have won a state lottery more than once. Winning one lottery does not exclude a person from winning other lotteries. As long as a person buys a ticket, his chances in the second lottery are the same as those of everyone else who buys a ticket.

To demonstrate that something is not just a coincidence, a verifiable causal link must be found between the two events. The burden of proof is to show that the two events are not a coincidence, not vice versa. Consequently, the first event must be shown to set up, or put into motion, events or conditions that logically allow, encourage, or make the second event occur.

The next step from simple coincidence is correlation. Correlation is when some kind of demonstrable relationship does exist between two events. When event A occurs, event B seems to also occur with some regularity. In fact, the correlation between A and B may provide predictive usefulness that can be tested and is repeatable.

Correlations may, indeed, indicate that there is a direct cause-and-effect relationship between two events. However, they may also indicate that the two events simply share a common factor. For example, a great increase in the number of traffic accidents in a single day usually correlates well with the number of collapses of agricultural outbuildings in the following 24-hour period. However, car accidents do not cause outbuildings to collapse on their own. The common factor is wet, heavy snow.

A heavy snow usually makes driving conditions hazardous. A higher than normal number of accidents then usually results on a day when a significant snowfall occurs. Agricultural outbuildings are not built to the same code standards as buildings occupied by people, and some are old and rickety. Consequently, roof collapses occur more frequently in agricultural outbuildings the day after a heavy snow.

One of my favorite correlations is between air conditioner failures and pregnancies. While one obviously does not cause the other, what they have in common is power outages. Low voltage, voltage fluctuations, and blackouts cause air conditioner compressor motors and control equipment to fail at a higher rate. Likewise, the same power trips, outages, and blackouts that leave people in the dark promote cuddling. In turn, the increase in cuddling results in a higher than normal number of births 9 months after the event. The New York blackouts that occurred variously on August 17, 1959, June 13, 1961, November 9, 1965 (the big one), July 13, 1977, and most recently August 15, 2003, all attest to this amusing correlation.

Unfortunately, coincidence and correlation are often mistaken for causation. This is sometimes called the *post hoc ergo propter hoc* fallacy. The Latin phrase literally means "after this therefore because of this." Because event, action, or condition B occurs after A, event, action, or condition A is presumed to have caused B.

Applying the Scientific Method to Determine a Root Cause

In a laboratory setting, it is usual to design experiments where the variable being studied is not obscured or complicated by other effects acting simultaneously. The variable is singled out to be free from other influences. Various experiments are then conducted to determine what occurs when the variable is changed. Numerous tests of the effects of changing the variable provide a statistical basis for concluding how the variable works, and predicting what will occur under other circumstances.

In this way, theoretically, any accident, failure, crime, or catastrophic event could be experimentally duplicated or reconstructed. The variables would simply be changed and combined until the right combination is found that faithfully reconstructs the event. When the actual event is experimentally duplicated, it might be said that the reconstruction of the failure event has been solved.

There are problems with this approach, however. Foremost is the fact that many accidents and failures are singular events. From considerations of cost, time, logistics, and safety, the event cannot be repeated over and over in different ways just so some engineer can play with the variables and make measurements until the outcome perfectly matches the event being assessed.

It can be argued, however, that if there is a large body of observational evidence and facts about a particular accident or failure, this is a suitable substitute for direct experimental data. The premise is that only the correct reconstruction hypothesis will account for all of the observations, and also be consistent with accepted scientific laws and knowledge.

An analogous example is the determination of an algebraic equation from a plot of points on a Cartesian plane. The more data points there are on the graph, the better the curve fit will be. Inductively, a large number of data points with excellent correspondence to a certain curve or equation would be proof that the fitted curve was equal to the original function that generated the data.

Thus, the scientific method, as it is applied to the reconstruction of accidents and failures, is as follows:

- A general working hypothesis is proposed based upon "first cut" verified information.

- As more information is gathered, the original working hypothesis is modified to encompass the growing body of observations.
- After a certain time, the working hypothesis could be tested by using it to predict the presence of evidence that may have not been obvious or was overlooked during the initial information gathering effort.

A hypothesis is considered a complete reconstruction when the following are satisfied:

- The hypothesis accounts for all the verified observations.
- When possible, the hypothesis accurately predicts the existence of additional evidence not previously known.
- The hypothesis is consistent with accepted scientific principles, knowledge, and methodologies.

The scientific method, as noted above, is not without some shortcomings. The reality of some types of failures and accidents is that the event itself may destroy evidence about itself. The fatigue fracture that may have caused a drive shaft to fail, for example, may be rubbed away by friction with another part after the failure has occurred. Because of this, the fatigue fracture itself may not be directly observable. Or perhaps the defective part responsible for the failure is lost or obscured from discovery in the accident debris. Cleanup or emergency repair activities may also inadvertently destroy or obscure important evidence. In short, there can be observational gaps.

Using the previous graph analogy, this is like having areas of the graph with no data points, or few data points. Of course, if the data points are too few, perhaps several curves might be fitted to the available data. For example, two points determine not only a simple line, but also an infinite number of polynomial curves and transcendental functions.

Thus, it is possible that the available observations can be explained by several hypotheses. Gaps or paucity in the observational data may not allow a unique solution. For this reason, two qualified and otherwise forthright experts can sometimes proffer two conflicting reconstruction hypotheses, both equally consistent with the available data.

The Scientific Method and the Legal System

Having several plausible explanations for an accident or failure may not necessarily be a disadvantage in our legal system. It can sometimes happen that an investigator does not have to know exactly what happened, but rather what did not happen.

In our adversarial legal system, one person or party is the accuser, prosecutor or plaintiff, and the other is the accused, or defendant. The plaintiff or prosecutor is required to prove that the defendant has done some wrong to him, his client, or the state. However, the defendant has merely to prove that he, himself, did not do the wrong nor has a part in it; he does not have to prove who else or what else did the wrong, although it often is advantageous for him to do so.

For example, suppose that a gas range company is being sued for a design defect that allegedly caused a house to burn down. As far as the gas range company is concerned, as long as it can prove that the range did not cause the fire, the gas range company likely has no further concern as to what did cause the fire. ("We don't know what caused the fire, but it wasn't our gas range that did it.") Likewise, if the plaintiffs cannot prove that the particular gas range caused the fire, they also may quickly lose interest in litigation. This is because there may be no one else for the plaintiffs to sue; there is insufficient evidence to identify any other specific causation.

Likewise, suppose that a crime has been committed and a person, a defendant, has been charged with the crime. If the defendant can simply prove that he did not do it or that it was not possible for him to have done it, that is sufficient. The defendant does not have to prove who actually committed the crime; he only has to prove that he did not do it.

Thus, even if the observational data are not sufficient to provide a unique reconstruction of the failure or crime, it may be sufficient to deny a particular one. That may be all that is needed.

Convergence of Independent Methods

The veracity of a scenario can be reinforced greatly if several independent methods are simultaneously used to reach the same conclusion.

For example, trajectory calculations that use physical evidence in combination with the known ballistic characteristics of the bullet and gun may show that some shots that were fired came from a certain location. Witnesses that heard the shots may further corroborate this finding by providing sound vectors that intersect in the area that coincides with the calculation. Examination of the area in question may find closely grouped bullet casings on the ground that match the bullet and the number of shots that were heard. Lastly, a witness may have observed someone shooting from the spot in question at the correct time.

Each of the above lines of evidence is wholly independent of the others, but each produces the same conclusion about the location of the shot. Some of them help support each other. Thus, the conclusion in the example about where the shots originated is not based upon just one line of inquiry, but several independent lines of inquiry. Each additional line of independent inquiry that logically supports or is consistent with the same conclusion further improves the probability of the veracity of the conclusion.

Occam's Razor

William of Occam was a medieval Franciscan monk and philosopher. He lived from 1285 to 1349 and studied theology at Oxford. Friar Occam would have received his master's degree from Oxford, but was denied the degree because the faculty, specifically the chancellor at the time, Jogh Lutterell, objected to some of William's ideas concerning apostolic poverty.

Being a Franciscan monk, William believed in living life as simply as possible. However, as occurred to many independent thinkers in those times, he ran afoul of church policy. Consequently, William was summoned to the papal court in 1327, which was then in Avignon, France. Charges had been brought against William by his old nemesis, Mr. Lutterell, again concerning his views on apostolic poverty. The chancellor was apparently still very steamed about William's opinions and wouldn't let go of the issue.

As an aside, I challenge the reader to research what this burning fourteenth-century issue, apostolic poverty, was all about. I use the word *burning* advisedly, because that was the usual punishment then for expressing a non-church-sanctioned opinion. Apparently, William's opinion about the matter was so dangerous to the civilized world that Mr. Lutterell was quite willing to ruin William's career and possibly even put him to death.

As the pope was about to issue a condemnation in 1328, Occam and two other Franciscans fled to the protection of Emperor Louis IV in Munich. William was eventually excommunicated from the church. However, not to be outdone, William authored a treatise that concluded Pope John XXII was a heretic.

Despite this relatively interesting life for a monk, Occam is chiefly remembered in the expression "Occam's razor." Occam's razor is not a principle or a law. It is a guide or rule of thumb. Sometimes called the principle of parsimony, or the principle of simplicity, in the original Latin writing, William of Occam wrote: *Pluralitas non est ponenda sine neccesitate*. Translated literally, this means: "plurality should not be put forward without necessity."

This means that when there are two or more scenarios that explain how an event may have occurred, the simplest scenario that fits the facts is usually the one that is correct. A more modern and succinct equivalent expression is: keep it simple, stupid. With respect to failure analysis, a person who applies Occam's razor:

- Cuts away redundant or unnecessary steps in the scenario hypothesis that cannot be proven or disproven

- Includes only those steps physically necessary to affect change from the initial condition to the final condition
- Uses the minimum number of steps to accomplish the change
- Leaves standing in the scenario only those steps that are consistent with the facts and proven science

The last point is sometimes called the coherence theory of truth. This theory asserts that a statement or scenario can be true only if it is logically consistent with other proven scientific theories and verifiable facts. A statement or scenario, therefore, cannot be true if it contradicts or is inconsistent with proven science or verifiable facts. As with Occam's razor, the coherence theory of truth is more of a guideline or rule of thumb. It seems to work most of the time but is not foolproof.

Here is an exaggerated example of the application of Occam's razor. One person might note that hot dogs can be cooked nicely over a charcoal fire. His explanation for how this occurs is that the coals glow and give off smoke. The smoke rises to the heavens. A god smells the smoke, which smells like a sacrifice. The god is pleased by the sacrifice and causes the hot dogs to be cooked. By applying Occam's razor, however, the scenario is much simplified: hot dogs placed over glowing hot coals become cooked. The imagined steps involving a god being pleased are not testable, provable, or needed to affect the change and, therefore, are deleted.

The fact that Occam's razor is a guideline does not mean that it does not have any scientific foundation. It does. Occam's razor is a disguised way of stating Le Chatelier's principle, the principle of least action, the principle that water flows in the direction of the steepest gradient, the principle that electricity follows the path of least resistance, the principle of diminishing returns, the second law of thermodynamics, Pareto's principle, and several other similar laws and principles.

By the way, Occam did not discover or invent the guideline himself. It was already a common logical device used in medieval philosophical argument. Occam, however, did use the logical device frequently in his writings. For that reason, the principle became associated with him.

Here is another way that Occam's razor can be applied. In accident or failure cases where there is no human intervention during the failure sequence of events, events unfold in a logical, step-by-step fashion. Each step will be a logical consequence of the conditions and actions existing in the preceding step. When sufficiently accurate information is available that describes the conditions and events in step 1, the application of scientific principles and methods allows a person to anticipate and predict the conditions and events that will then exist in step 2, and then likewise in step 3, step 4, and so on. Without interference, naturally occurring events follow the rules of nature.

However, when there is human intervention in the scenario, often at least one step in the sequence does not proceed logically from the preceding step. The events may proceed logically from the initial condition to the point of human intervention, and then proceed logically after the point of human intervention to the final condition. At the point of human intervention, however, the apparent human-controlled action is illogical or out of sequence; that is, it is inconsistent with naturally occurring events in a closed system.

In arson, for example, by definition the arsonist arranges how, when, and where the fire occurs. While he may try to make the arrangement appear natural, conceal his handiwork so that it won't be discovered, or perhaps discretely "help" conditions already present that are conducive for a fire, the arsonist supplies the missing ingredient that heretofore had prevented the fire from occurring spontaneously.

In working backwards through the fire timeline evidence, from final condition to initial condition, a person may find that the failure sequence, that is, the fire in this case, is logical and orderly until a particular unnatural step is reached. The unnatural step is likely the point where human intervention occurred to direct, magnify, or initiate the fire. At that point, Occam's razor can be applied to hypothesize the simplest, easiest action that accounts for the unnatural step.

With respect to the above, a physicist or engineer would say that natural events proceed according to the principle of least energy. That is, nature favors the most energy-efficient path between two points, conditions, or states. Similarly, a chemist would say that events proceed according to La Chatelier's principle, a variant of the principle of least energy. Further, when events occur spontaneously, a chemist would say that the change in Gibb's free energy for the system is less than zero or that the system's entropy has increased.

Consequently, when an event or a series of connected events do not follow a course consistent with the principle of least energy or its corollaries, it is likely that something external to the system boundary has affected or initiated the course of events. To an investigator, this is a good place to begin looking for possible human intervention.

Organizing Evidence 4

Errors, like straws, upon the surface flow;
He who would search for pearls must dive below.

John Dryden, from "All for Love," 1768

Data Collection and Efficient Sorting Schemes

As was previously discussed in Chapter 1, in the beginning of an investigation the available facts and information are like pieces of a puzzle found scattered about the floor: a piece here, a piece there, and one piece perhaps that has mysteriously slid under the refrigerator and won't be found until after the puzzle has been solved.

At first, the puzzle pieces are simply sought out, gathered, and placed in a single heap on a table. Then, the pieces are sorted into smaller heaps. Perhaps all the pieces with straight edges are put into one pile, while pieces with a common color or textural appearance are put into other piles.

Pieces within these piles are fitted up to each other until a few pieces are found that fit together. If someone has a notion of what the puzzle picture might be, the fitted pieces are then laid out approximately where they might go. For example, green-colored groups of fitted pieces might be tentatively placed at the bottom, where there is supposed to be grass, and blue-colored groups of fitted pieces might be placed near the top, where the sky is expected to be. The small groups of fitted pieces are initially like disconnected islands surrounded by a large amount of empty space.

As more pieces are fitted together, the puzzle picture begins to emerge. The islands consisting of just a few fitted pieces become larger and merge with other islands. As portions of the picture develop, it becomes easier to figure out where the remaining puzzle pieces fit. Eventually, when all the pieces are fitted together, except for the one still hidden under the refrigerator, the puzzle is solved and the picture is plain to see.

Efficiently solving a picture puzzle and efficiently solving a root cause investigation have several areas in common. Initially, all the pieces that belong to the puzzle have to be found and collected. If there are missing pieces, this diminishes the ability to form the whole picture. Likewise, if pieces are included that belong to another puzzle, this can cause confusion until it is recognized that the errant pieces do not belong to this puzzle.

When puzzle pieces are collected, they are sorted into groups, which is sometimes called binning, according to a reasonable scheme. Binning aids in efficiently fitting the puzzle pieces together. In a thousand-piece puzzle, it is easier and faster to determine how a related group of one hundred pieces might fit together than it is to try all one thousand pieces at once.

Consider the following example. Suppose a person randomly takes one puzzle piece from a thousand-piece puzzle. This puzzle piece has four places on it that can be matched to four other puzzle pieces. While it might be necessary to check all 999 remaining pieces to find the four matching pieces, a person may have to try perhaps only 750 pieces to find the four matching pieces. If he is successful in finding the four matches, he now has clump of 5 solved puzzle pieces and 995 remaining pieces.

If there are no edge pieces, the assembled clump of five puzzle pieces will have eight places that allow eight new pieces to be fitted into the clump. As before, while it might be necessary to check all 995 remaining pieces, the person will likely have to check perhaps only 870 pieces to find the 8 matching ones. Now the person has 13 pieces of the puzzle solved.

If there are still no edge pieces in the clump, the clump of 13 puzzle pieces will have 12 places that will allow 12 new pieces to be fitted into the clump. Using the same reasoning as before, the person will now likely have to try about 905 pieces to find the 12 matching ones. Now the person has 25 pieces of the puzzle solved.

With 25 pieces, the person now has 16 open places that will allow 16 more pieces to be fitted into the clump. With 975 pieces left, he will, on average, have to make 914 tries to fill in these positions with matching pieces. If successful, he will then have 41 pieces of the puzzle solved, and 959 pieces remaining.

In other words, to solve the puzzle for just 41 pieces, or 4% of the puzzle, using this simple but brute force method, the person has had to make around 750 + 870 + 905 + 914 = 3,439 fit-up tries with all the puzzle pieces. However, there are still 959 more pieces left to fit into the puzzle. While this method is straightforward and logical, it is indeed a daunting way to solve the puzzle.

Alternately, let's say that the person who wants to solve the puzzle divides the puzzle into ten piles. One pile might be all edge pieces, another might be all pieces that are sky blue, another might contain pieces that appear to be mountain along the horizon, and so on. In other words, the pieces are divided into piles that appear to share commonalities in the puzzle. Now, assume that each pile is composed of one hundred pieces that will all fit up to each other. If our patient puzzle solver selects one piece at random from a particular pile, he will have to make about 75 tries to assemble the first five pieces into a clump. To assemble 13 pieces, he will need to make 88 more tries. To assemble 25 pieces, he will have to make about 44 more tries. And finally, to assemble 41 pieces, he will need to make perhaps 38 more tries. This is a total of 75 + 88 + 44 + 38 = 245 tries.

In sum, using the simple brute force method with no initial systematic sorting of the puzzle pieces could take about 3,439 fit-up tries to assemble 41 pieces of a thousand-piece puzzle. On the other hand, by systematically sorting the puzzle pieces into ten equal piles, the same result of 41 puzzles pieces being assembled can be accomplished with only about 245 fit-up tries. This is a reduction in fit-up tries of 93%.

While the brute force method is logical and simple, it also is tedious, is time-consuming, and consumes resources prodigiously. The binning method significantly reduces the amount of work needed to solve a puzzle, but it also requires a logical, systematic binning scheme.

Systematic binning of puzzle pieces has other advantages. Extraneous puzzle pieces that do not belong to this particular puzzle, which somehow got into this puzzle pile by mistake, are generally easier to detect during the binning process. Further, binning allows the puzzle to be solved by several groups of people working in parallel at the same time. When each group of people has its pile of pieces solved, each group can then match up their finished clump of assembled puzzle pieces with the other groups' clumps. A picture will then start to emerge.

In the example cited previously, if each pile of 100 pieces were assembled in just 245 fit-up attempts, putting together the ten large clumps only requires a few more fit-up attempts, perhaps nine or ten. This means that the whole puzzle can be solved in about 3,018 fit-ups. Without binning, it took that same number of fit-ups just to solve 4% of the puzzle.

Making sense of all the facts collected during an investigation is not much different from making sense of a pile of puzzle pieces.

Verification of Facts

The scientific investigation and analysis of an accident, catastrophic event, failure, or crime is structured like a pyramid. There should be a large foundation of verifiable facts and evidence at the bottom. These facts and evidence support the analysis that is performed in accordance with proven scientific principles and logic. The facts, evidence, and analysis, taken together, then support a small number of conclusions that form the apex of the pyramid.

Conclusions should be directly based upon verifiable facts and analysis. They should not be based upon other conclusions, assumptions, or hypotheses. If the facts are arranged logically and systematically, the conclusions should almost be self-evident. Conclusions that are based upon other conclusions, assumptions, conjectures, or hypotheses, that in turn are based only upon a few selected facts and generalized principles, however, are a house of cards. When one point is proven wrong, the entire complicated logical construct collapses.

Consider the following. As noted in Chapter 2, in most investigations the application of cogent arguments is more typical than the application of airtight deductive arguments. Consider what occurs when a cogent argument is true 90% of the time.

If the foundation of an argument depends upon ten verified facts and there are no facts that falsify the argument, and each fact independently implies that A is true with a probability of 90%, then the chance that the final conclusion is true is $[1 - (1 - 0.90)^{10}] = 99.99999999\%$.

However, if the foundation of the argument is supported by just one fact, there are no facts that falsify the argument, and from that one supporting fact a series of ten conclusions are made, each conclusion depending upon the previous one, then the final conclusion has a probability of being true only 0.90^{10} or 34.9% of the time.

Further, if one of the supporting ten facts in the first instance was not verifiable, this would reduce the probability that the conclusion reached with the other nine facts is true 99.9999999% of the time. However, if the one fact supporting the argument in the second instance is not verifiable, the probability of the final conclusion being true is zero.

The following are some generalized guidelines for ensuring that the facts are accurate.

- Whenever possible, use primary sources. Primary sources are verifiable facts. Primary sources are not opinions, beliefs, guesses, assumptions, presumptions, or conclusions. Primary sources may be photographs of the evidence, sound or video recordings of the event, firsthand observational records of the event in logs, data recorded on output charts or printouts of data automatically logged during the event, records of observations by credible witnesses of the event, and samples of materials or physical evidence taken from the place where the event occurred. Perhaps a better name for a primary source is a firsthand source.

  It is good practice to note the location or origin of primary sources used in an investigation not only so that they can be found again if needed for demonstrative purpose, but also so that their veracity can be double-checked by other investigators. If a primary source cannot be retained for long, attempt to document it and its relevant characteristics in case its validity is challenged later. Having impartial witnesses present during the documentation who can vouch that the documentation faithfully represents the condition of the item at that time is desirable.

- When secondary sources are used, that is, materials developed, compiled, and reported by other people or that otherwise are one step removed from the primary sources, be sure to note the particulars of where the information can be found, who provided it, and so on. Since secondary sources can run the gamut from wholly unreliable to rock solid reliable, it is best to know the reputation of the secondary source. Obviously, if the information is important to the conclusions of the investigation, it is best to use only secondary sources that have been proven to be reliable.
- Whenever possible, use several independent sources to verify a fact. A well-constructed sophistry can appear to trump or explain away a single fact. A fact that can be verified by several independent, well-regarded sources, however, is difficult to explain away by sophistry.
- Facts that are fundamental to the conclusions of an investigation should be documented. This is considered due diligence for an investigator. Due diligence is the measure of assiduity ordinarily expected from a reasonable and prudent investigator to ensure that the facts upon which an investigation is based are accurate and verifiable.

  What constitutes due diligence is subjective and varies with the situation and type of investigation being done. However, one technique is to imagine yourself on the other side of the investigation. What facts would you point to as being unreliable, suspect, or incorrect if you were being accused of a crime, malfeasance, or something similar? In that sense, those are exactly the facts that must be checked, verified, and documented.
- Creating a file or notebook for all the notes and hardcopy items important to the investigation is very useful. In some cases, if information is being held in computer databases or files, hardcopy backups can save a lot of grief when a computer crashes or files are accidentally deleted. With respect to backup copies of computer databases, keep the backup copies in a completely different location to avoid both copies being stolen at the same time, burned up at the same time, and so on. This is called a common cause failure.
- Be cognizant of proprietary information, confidential information, information related to security, personal information, and so on. Be sure to provide standard acknowledgments when using copyrighted materials. While it may not be required by law, common courtesy dictates proper acknowledgment of other people's work or research when appropriate.
- When peer reviews are important, keep records of who reviewed what, and when it was done. When new facts are discovered that significantly change conclusions, these dates can become important in establishing a record of due diligence.

- Adhere to appropriate chain of custody protocols to preserve the integrity of evidence. A chain of custody protocol refers to the documentation, procedures, and methods used to ensure that the integrity and condition of evidence is preserved, and that it is not altered during its transfer, seizure, movement, examination, or analysis. Evidence should be handled properly to not only prevent damage, alteration, degradation, and contamination, but also to protect it from tampering and substitution. If physical evidence is altered, independent parties cannot verify the facts that were learned or asserted from the initial examination of this evidence. In cases involving legal prosecution, this can significantly compromise the case. Further, as was discussed in Chapter 2, evidence will often be examined two or more times to look for items or characteristics not originally considered during the first examination.

  A good discussion of chain of custody legal requirements for the state of Delaware can be found at the following website: http://www.delcode.state.de.us/title10/c043/sc03/. While perhaps not exactly the same for each state or U.S. territory, these requirements are similar to standards used in most state and federal courts.

Organization of Data and Evidence: Timelines

A good place to begin sorting data and evidence is with the creation of a timeline. A timeline is a chronological arrangement of the events associated with the incident, crime, or failure being investigated. Timelines can be as simple as a single-column ordered list of events indicating what came first, what came second, and so on. Such a timeline can be conveniently done on a spreadsheet in tabular format. Table 4.1 gives an example of a timeline that involves events associated with failure in an emergency diesel generator unit.

More sophisticated timelines can be timescaled, like a construction project Gantt chart schedule. By adding a timescale the relative duration of events and the time between events are proportional to the actual time intervals. Table 4.2 is an example of an investigation Gantt chart that was used to plan a failure investigation at the beginning of the investigation.

Table 4.1 Example: Spreadsheet Event Timeline

Time	Sensor Point	Remarks
13:03:30.425	3917	4160 v bus 1G bkr 1GE lockout
13:03:30.529	3650	Diesel gen 2 bkr EG2 trip
13:03:30.529	3911	4160 v bus 1G bkr 1GE trip
13:03:35.248	3659	Diesel gen 2 lockout

Table 4.2 Example: Gantt Chart Investigation Plan Timeline

ID	CR-CNS-2007-00480	Start	Finish	Jan 21 2007	Jan 28 2007	Feb 4 2007	Feb 11 2007	Feb 18 2007
				23 24 25 26 27	28 29 30 31 1 2 3	4 5 6 7 8 9 10	11 12 13 14 15 16 17	18 19 20 21 22
1	Event occurred, LCO 3.8.1	1/18/2007	1/22/2007					
2	Problem statement	1/23/2007	1/26/2007					
3	Evidence gathering	1/18/2007	2/2/2007					
4	Analysis of evidence	1/18/2007	2/7/2007					
5	Preparation of RC Report	1/24/2007	2/22/2007					
6	Development of CA's	2/19/2007	2/22/2007					
7	1 Week update	1/31/2007	2/1/2007					
8	2 Week update	2/6/2007	2/8/2007					
9	3 Week update	2/13/2007	2/15/2007					
10	Presentation to CARB	2/22/2007	2/22/2007					
11	Final approval of report	2/22/2007	2/23/2007					

Similarly, Figure 4.1 is an example of another kind of timeline. This one was also used to plan a failure investigation. A similar type of timeline can be laid out on a blackboard using Post-it notes to denote information and events. The Post-it notes are handy because they can be easily moved around and added to when needed. Like the Gantt chart, the example in Figure 4.1 can use brackets to denote items that occur over a period of time.

In some cases, it might be useful to have several timelines set up in parallel to each other. For example, in a power plant turbine accident, the events or conditions occurring in the turbine system might be on one timescaled timeline. Alongside this might be another timeline showing what was going on in the control room. Both timelines would be matched to the same timescale so that what was happening in the turbine could be easily compared with what was going on in the control room at the same time. A third timeline might then be added to show what was going on with the output of the generator, a fourth might be added to show what else was going on in the plant at the same time, and so on.

With respect to a crime, sabotage, or destructive act, timelines are very useful in assessing potential suspects. If the time sequence of the event is reasonably known and laid out to scale, the verifiable activities of the various suspects can be similarly laid out in parallel to it. This will make plain which of the initial suspects had the right amount of unaccounted for time, at the right time, to have done the deed.

The use of such a timeline can often rapidly reduce the number of prospective suspects. While many suspects may have wanted the victim to die or had a reason to be angry at the company, only perhaps a handful might have had the available time at the right time to do the deed.

A timeline can also be correlated to a plot of a person's activities on a map or scaled drawing. While a timeline might show if a person had the available

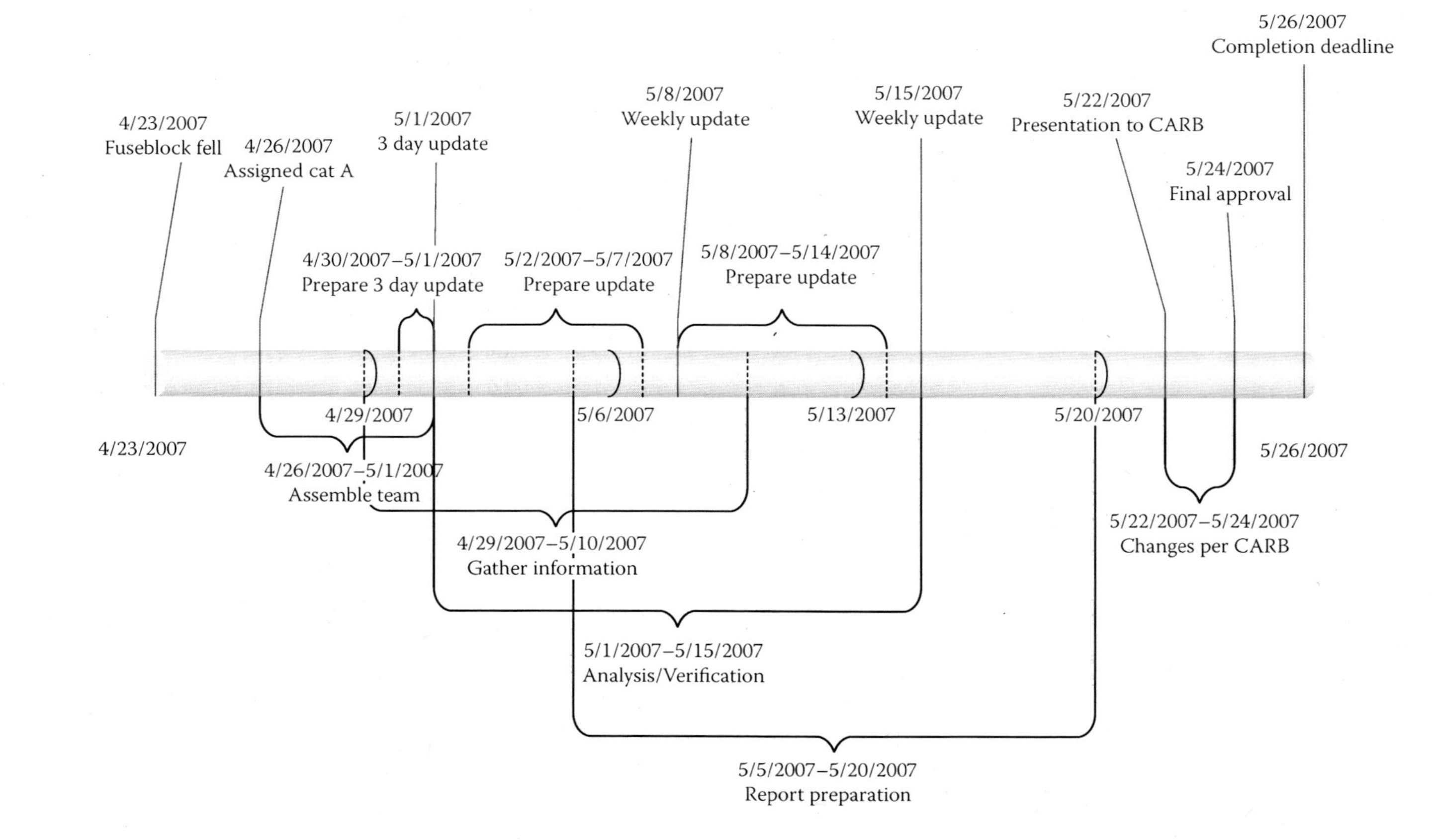

Figure 4.1 Example: Scaled timeline with notes and durations.

time at the right time to do the deed, a plot of his verifiable positions at the various times on the timeline might further provide information as to whether he was in the right location, or could have been, to accomplish the deed.

Cause-and-Effect Diagrams

Presume that a tank of compressed air has ruptured and several people have been injured and several pieces of equipment have been damaged. A simple initial cause-and-effect diagram of the failure might look something like Figure 4.2. The immediate cause of the effect, that is, the damages and injuries, in this case is obvious: The compressed air tank ruptured. The evidence to support this includes eyewitness accounts, the condition of the tank itself, and the matching of missile fragments to the tank.

The next level of the diagram might look like Figure 4.3. In Figure 4.3 the immediately preceding possible causes for the tank to rupture are considered. Note that the five level 2 categories for this example are kept as general as possible. In a sense, the branching at each level of a cause–effect diagram is similar to the methodology applied in biological taxonomic systems or cladistics. By doing so, whole branches of failure possibilities can sometimes be readily eliminated by the available evidence without having to expend a great deal of time or resources.

For example, if it can be shown that at no time was there a loss of electrical power in the system, then all the failure scenarios involving the failure of an electrical component can be eliminated.

A cause-and-effect diagram lends itself well to the use of cogent or conditional deductive reasoning. Sometimes there is historical or heuristic data available that allow probabilities to be assigned to the various possible causes.

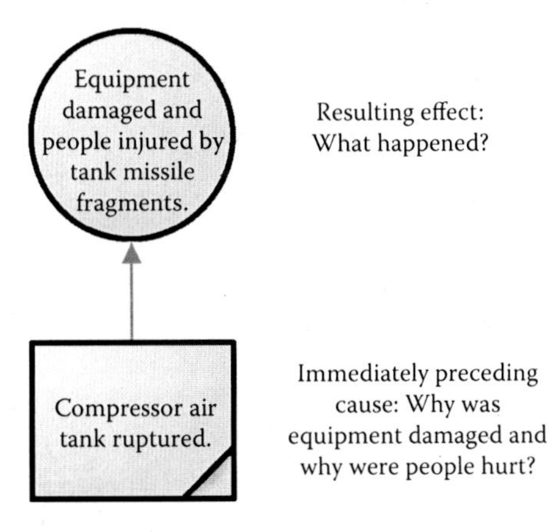

Figure 4.2 Initial cause-and-effect diagram.

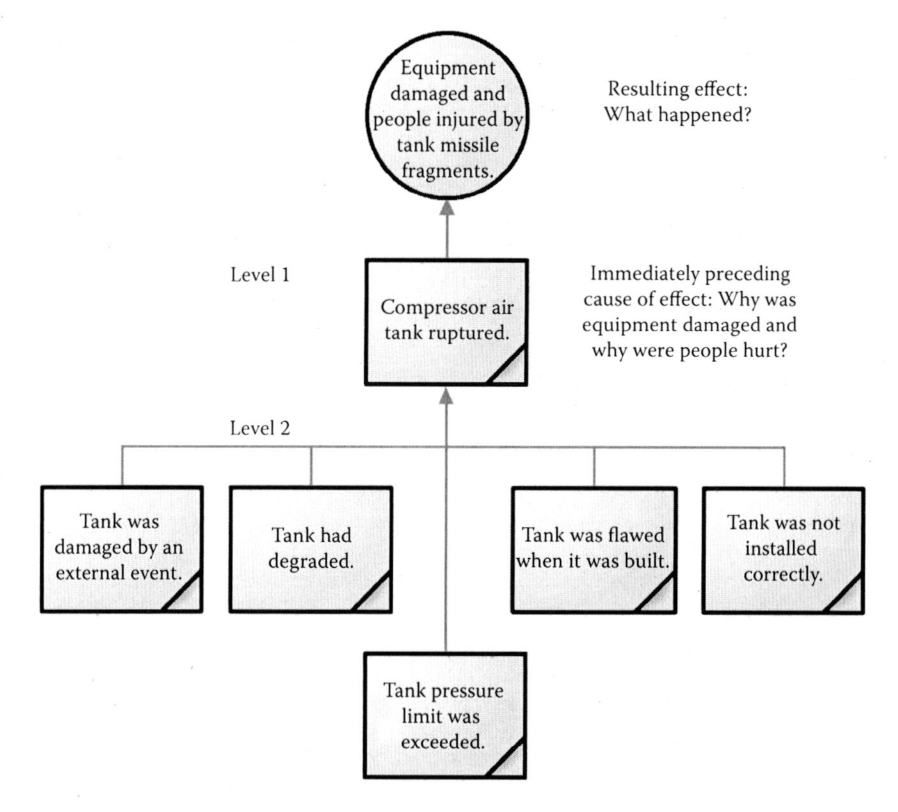

Figure 4.3 Cause-and-effect diagram: second level of possibilities.

In this case at hand, for example, if the compressor and tank had been in place for 10 years, would it be reasonable to conclude that there is high probability that the unit was initially installed correctly and that there were no pertinent internal flaws? Such a conclusion might be reasonable if experience has shown, for example, that 99% of all compressor installation errors are detected during the first 6 months of operation, or that 95% of all manufacturing flaws are found during the first month of continuous operation, which is sometimes called the infant mortality operational period. Likewise, the same historical or heuristic data might indicate that if the compressor and tank had been in place and operating for 10 years, some kind of age-related effect might be indicated.

Some gathering of simple background information and an examination of the event area might readily eliminate the possibility that the tank was damaged by some kind of external event, such as being impacted by an object, sabotage, being inadvertently damaged by nearby welding work, or perhaps external corrosion by exposure to chemicals. Likewise, verification that the relief valve was in good condition and confirmation by data logs or recorders that tank and system pressure limits were not exceeded would readily eliminate the possibility that the tank was excessively pressurized.

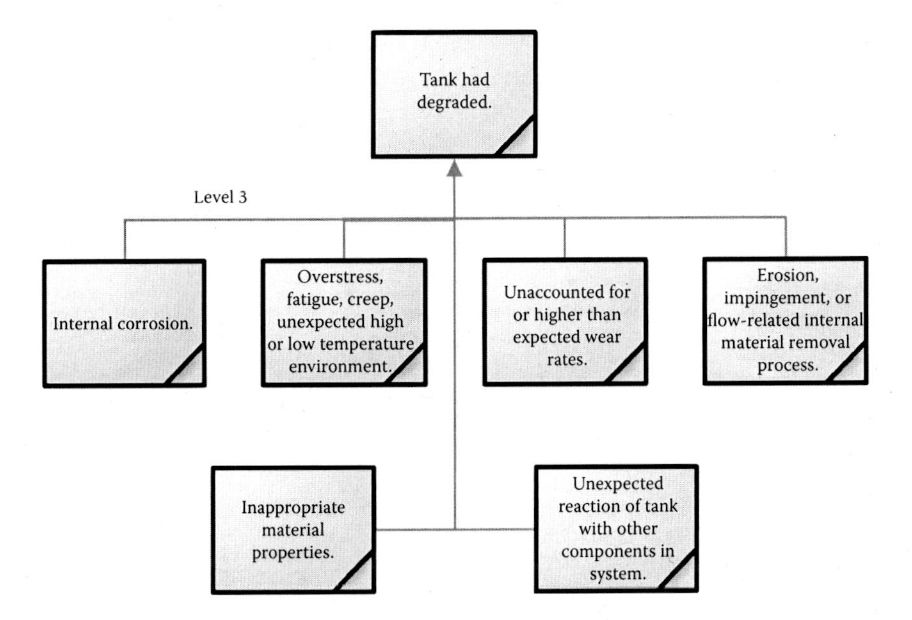

Figure 4.4 Cause-and-effect diagram: third level of possibilities.

Presuming that the evidence cannot falsify that the tank somehow had degraded over time, but does falsify the other general possibilities, a level 3 diagram concerning tank degradation could be developed as shown in Figure 4.4. By proceeding methodically in this way, following the evidence and comparing it to hypothesized failure modes, the failure mode cause-and-effect relationships can be identified.

Figure 4.5 an example of a slightly different type of cause-and-effect diagram that utilizes OR gate logic flowchart symbols. It depicts possible reasons why a relief valve unexpectedly opened and allowed pressurized hot water to escape. The escaping pressurized hot water immediately flashed into steam and then condensed. This caused habitability problems in the adjacent plant area, which then required the plant to reduce power to fix the problem. An unscheduled reduction of power at a base load power plant has obvious economic consequences as well as administrative ones.

The first level of possibilities (Figure 4.5) shows the final, undesirable effect or consequence in the circle. Note that the first three possibilities are very general: it operated as designed, it malfunctioned, or something or someone caused it to open. A brief reason why two of the three possibilities were eliminated from consideration is noted under their respective boxes, and the diagonal lines through the boxes indicate that the available evidence was sufficient to eliminate them from further consideration. The note under the third box indicates that the evidence shows it did not operate according to its design specifications.

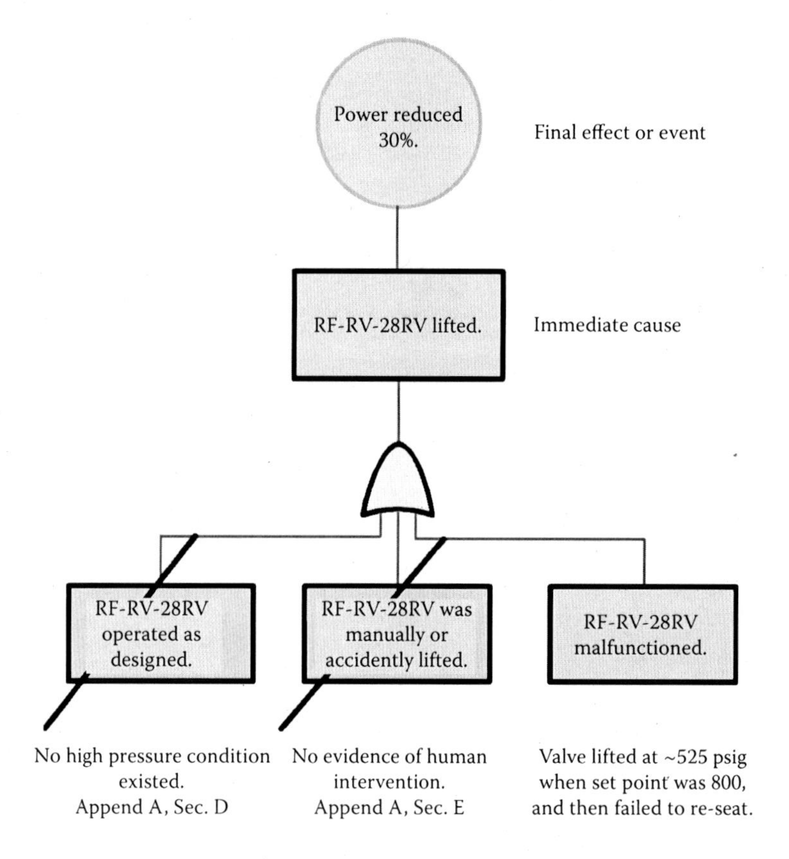

Figure 4.5 First-level cause-and-effect diagram for relief valve that unexpectedly lifted. RF, reactor feed water; RV, relief valve. RF-RV-28RV is simply the designation of a specific relief valve.

The next cause-and-effect level is shown in Figure 4.6, where the possible reasons why the relief valve might have malfunctioned are considered: damage during installation, a manufacturing defect, incorrect set points, and age-related degradation. As the diagram shows, three of the four possibilities were eliminated by the evidence, and the remaining one is validated by the evidence.

The next level of evaluation, shown in Figure 4.7, indicates in the dotted line to the side note or explanatory balloon that a diagnosis guide for relief valves published by the Electric Power Research Institute was consulted. An examination of the valve that failed was compared to the characteristic symptoms, or modes, of thirty-four types of failures. After a thorough examination of the valve in question, it was found that thirty-two of the listed applicable failure types did not fit the physical characteristics in this case. Only two items were found applicable, and they were related to each other, as shown in the diagram.

Figure 4.8 depicts the administrative steps that were not done or were missed that allowed the relief valve to become too old and then fail.

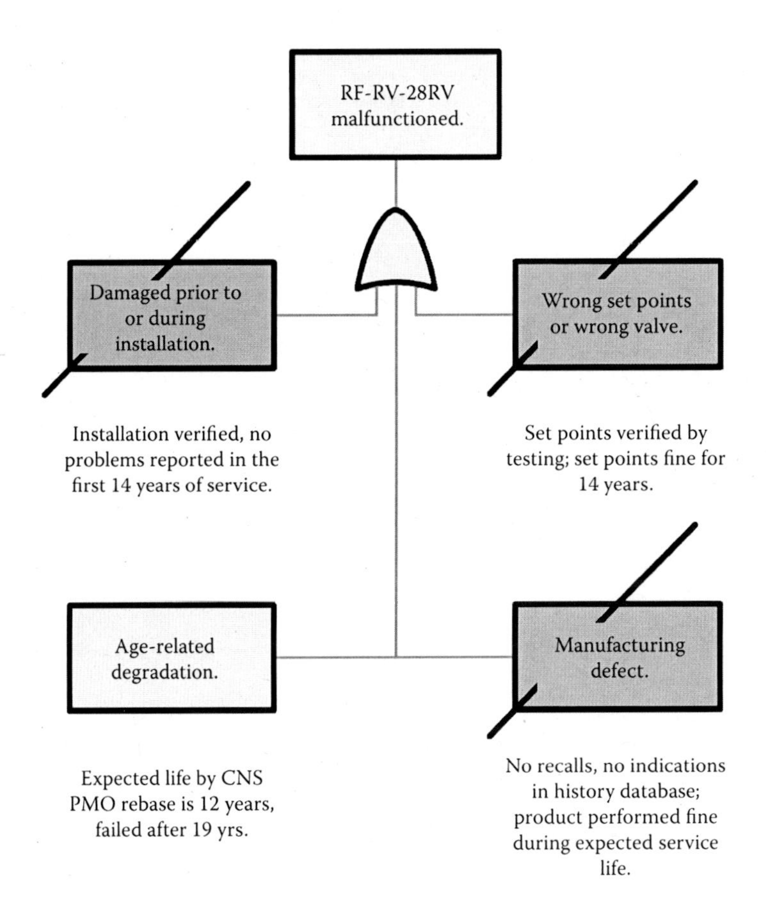

Figure 4.6 Second-level cause-and-effect diagram for relief valve that lifted.

Figure 4.8 begins (top box, the resulting effect) where Figures 4.6 and 4.7 left off: with the fact that the valve had degraded due to aging effects. The chain of administrative events or processes diagrammed below the top box in Figure 4.8 indicates that the relief valve was inadvertently not classified correctly in the preventive maintenance program. Consequently, preventive maintenance tasks (PMs) were not done on the valve to keep it in good condition.

The diagram boxes in Figure 4.8 also indicate that when there had been an opportunity to do preventive maintenance work on the valve, there were insufficient spare parts to do a good job, although it did pass its functionality test. Again, because it was incorrectly considered to be a low-priority component, no follow-up maintenance work was done.

The findings depicted in Figure 4.8 can be emphasized by cross-correlating them with a timeline, as shown in Figure 4.9. The scaled graphic comparison in the timeline of the standard replacement time to the expected replacement time and to the time that the valve was actually in service convincingly drives home the reason why the component failed.

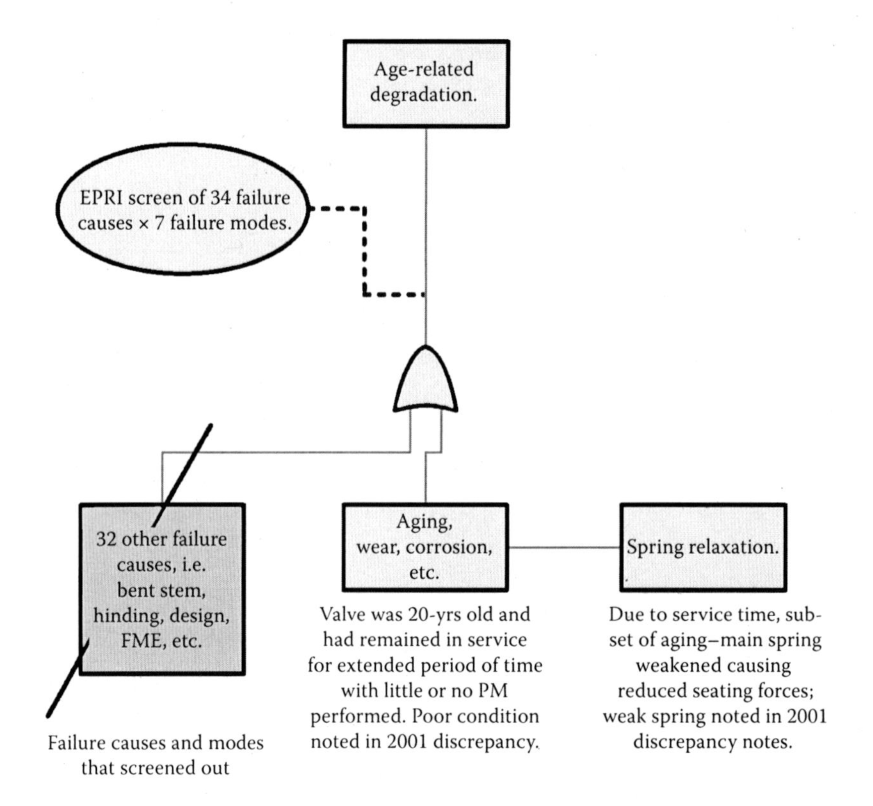

Figure 4.7 Third-level cause-and-effect diagram for relief valve that lifted. FME, foreign material in the valve that caused problems.

In essence, a cause-and-effect diagram, such as those shown in the previous examples, is similar to a failure modes and effects analysis (FMEA) diagram, except that it is constructed in reverse. A failure modes and effects analysis is a method used to identify, define, and examine potential problems, errors, or failures in a product, component, system, or process usually during the design process. It is a designer's best guess, perhaps updated with actual field experience, as to all the ways in which the product, component, or system can fail. As such, a good FMEA can provide the following information:

- The known ways in which the item can fail
- The consequences of a particular failure mode
- The specific characteristics of a failure mode, i.e., its specific failure signature
- The various causes and factors that contribute to a failure mode
- Failure rates and failure probabilities of a failure mode

A designer uses an FMEA to anticipate the various ways in which his product, component, or process might fail, and then develops contingencies

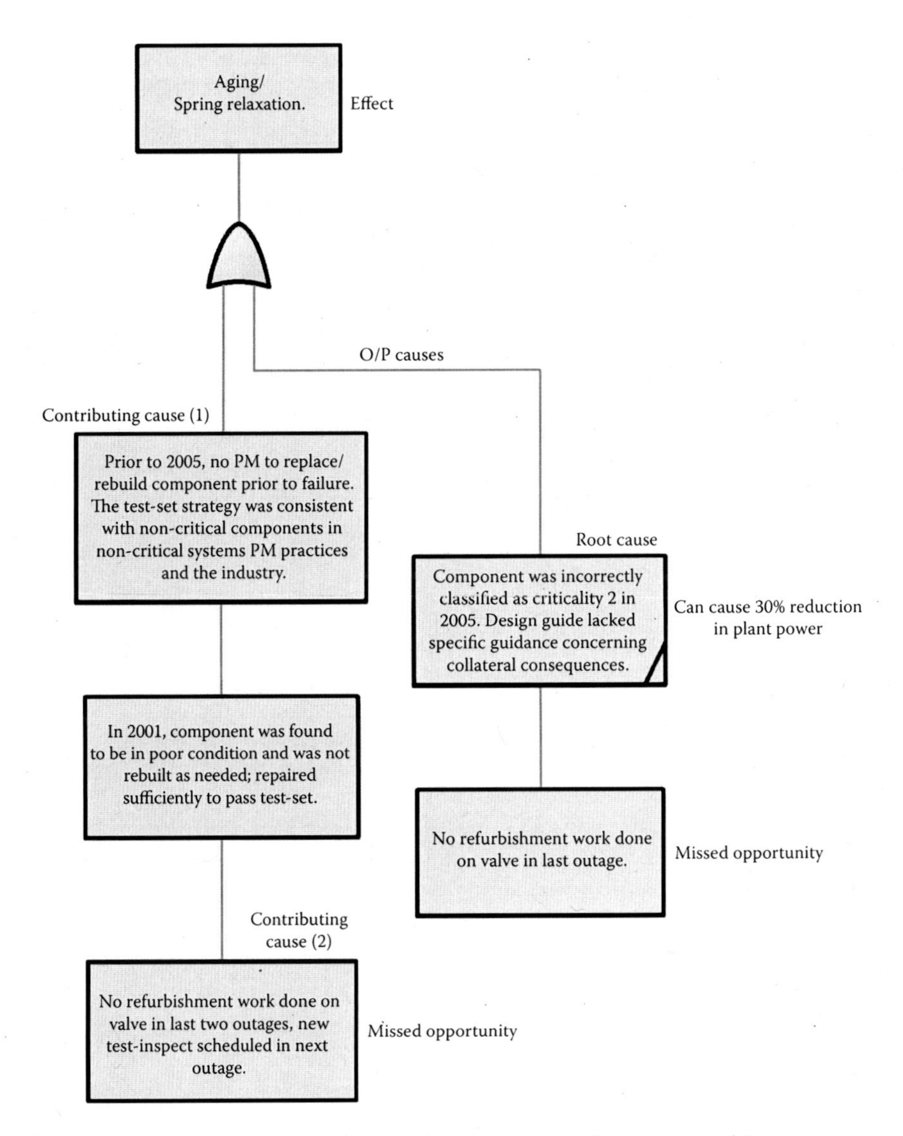

Figure 4.8 Fourth-level cause-and-effect diagram showing the administrative causative factors.

to eliminate, mitigate, or cope with the failure. As such, an FMEA is a long list of "what if's" as to what could go wrong and (hopefully) what to do about it.

Sometimes a reverse FMEA is thinly disguised in a vendor manual as a troubleshooting guide, or in publications that quantify the pattern of failures noted to occur in a particular type of machine, component, or system. Examples of such publications include the *IEEE Standard 500 Reliability Handbook*, *EPRI TR-105872 Safety and Relief Valve Testing and Maintenance Guide*, *Machinery Failure Analysis and Troubleshooting* by Bloch and Geitner (see entry in references chapter), and many others. There is a gold mine of

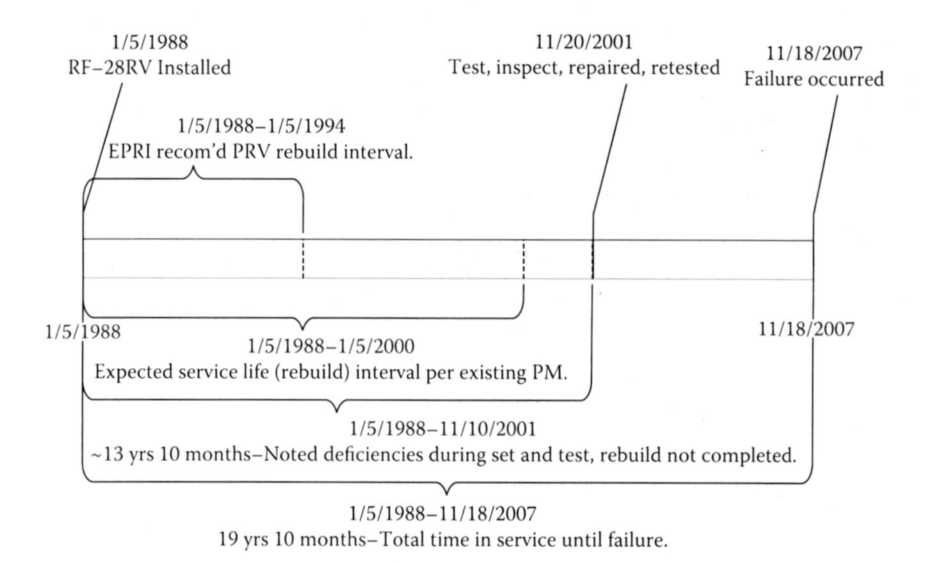

Figure 4.9 Timeline emphasizing the difference between the expected service life and the actual time in service.

practical failure experience in these publications that can be a great help in determining the cause of a component or system failure.

A Place to Start

After setting up a timeline, in lieu of not having any particular idea or notion of where to begin assessing the facts and evidence associated with a particular failure, one way to organize the data being collected is to bin it according to known failure modes and their associated probabilities.

Table 4.3 is an example that involves centrifugal pumps. The listed failure probabilities are based upon centrifugal pump failures that occurred in U.S. process plants. Data concerning such a centrifugal pump failure can be collected and organized according to how they relate to these seven failure mode causes.

It is sometimes useful to add one extra bin to account for data and information that does not appear to readily fit into any of the other categories. This helps prevent information from being forced into a bin and protects against deciding ahead of time that the failure must be one of the seven modes listed (an *a priori* assumption). This list of potential causes would then become the first level of a cause-and-effect diagram, such as shown in Figures 4.2 and 4.5.

Because the first two causes in the list, lack of maintenance and installation error, constitute about 55% of the failures that regularly occur in this equipment type, it might be efficient with respect to the allocation of investigative resources to begin collecting information first about the maintenance

Table 4.3 Failure Mode Probabilities of Centrifugal Pumps

Cause	Probability
Lack of maintenance, deficient maintenance	30%
Installation deficiencies, errors, defects	25%
Unexpected service conditions, incorrect design fit	15%
Improper operation	12%
Fabrication deficiencies	8%
Deficient design	6%
Material flaws, defects, deficiencies	4%
Unknown failure cause	na

Source: Bloch, H., and Geitner, F., *Machinery Failure Analysis and Troubleshooting*, 3rd ed. (Houston, Texas, Gulf Publishing, 1997), Table 10-1, p. 575.

history and installation of the particular centrifugal pump. As the accumulated evidence either supports or falsifies these potential causes, then the less probable causes can be pursued.

In that regard, an evidence-hypothesis table can be created to keep track of which hypotheses a particular fact or piece of evidence supports and which ones it falsifies. It might look something like that shown in Table 4.4. The potential causes along the top row of Table 4.4 are the same as those listed in Table 4.3.

As verifiable facts and evidence are collected, those items that support a particular hypothesis can be filled in with an X, and those items that falsify a particular hypothesis can be filled in with an O. Obviously, a box filled in with an O effectively kills that hypothesis.

Event and Causal Factors Diagrams

A timeline and a cause-and-effect diagram can be combined into a single graphic called an event and causal factor diagram. The sequence of events is essentially laid out linearly like a timeline. However, causal factors and conditions are diagrammed and connected to the events to show their cause-and-effect relationships.

The usual geometric symbols used in an event causal factors diagram are as follows:

- Validated events are typically represented by rectangles.
- Events that must have occurred, or which have been hypothesized to have occurred but which are not presently validated, are represented by rectangles drawn with dotted lines.

Table 4.4 Evidence Hypothesis Table

Evidence/Facts	Maintenance	Installation	Service Condition	Operation	Fabrication	Material Flaw	Unknown
Dimensions checked and correct		O			O		
All maintenance done was per vendor manual and on schedule	O						
All replacement parts receipt inspected and checked	O				O	O	
No fractures, breaks, or material problems						O	
Installed by factory representative and then tested		O	O				
Pump model has good record in industry					O	O	
Service condition matches design specifications			O				
Supply to pilot valve needs 200 gpm for start; low-pressure supply conditions only supplied 100 gpm recently				X			
Vendor manual warns about damage due to low pilot supply on start-up				X			
Damage matches that expected if pilot supply is low				X			

- Conditions presumed to have existed, but which have yet to be validated are drawn with dotted curves.
- A diamond shape represents a significant event in which a limit is exceeded, a failure has occurred, or some other bad event has happened.
- An ellipse represents a causal factor and is connected to the event it affects by an arrow. A root cause might be further denoted by a vertical line on the right side of the ellipse.
- The undesirable final event is usually represented by a circle.
- Time intervals are denoted by brackets.

Figure 4.10 is an example of an event and causal factor diagram that also contains a barrier diagram along the bottom. The barrier diagram indicates what work practices, rules, regulations, and so on were omitted, subverted, or otherwise not done and might have prevented the event from occurring.

In the example, the event being depicted is a failure in a small section of heat trace wiring in a nuclear power plant safety system. The heat trace was used to maintain a warm temperature, within certain high and low limits, in piping associated with a tank that contained a special chemical solution. The

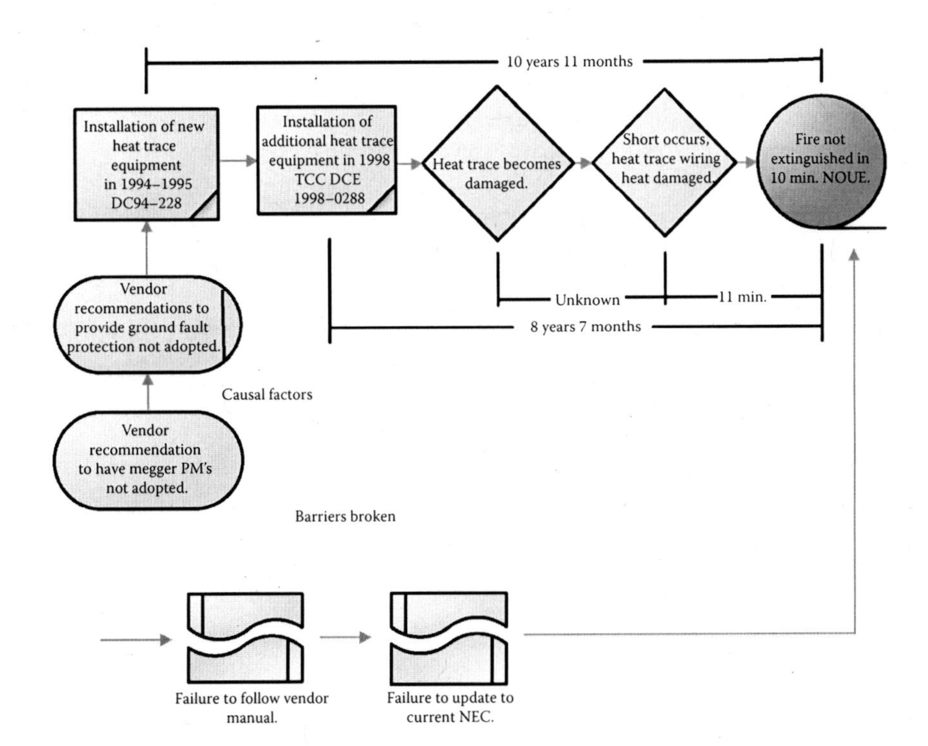

Figure 4.10 DCE, design configuration evaluation; TCC, temporary configuration change.

tank, which contains about 4,500 gallons of solution, was part of an emergency safety system called the standby liquid control (SLC) system. When needed, injection of the solution from the tank into the reactor would bring the reactor from a hot condition to a cold, shutdown condition. The solution inside the tank, a mixture of water and sodium pentaborate, absorbs neutrons and stops the reaction in the reactor.

What occurred is this. A portion of the heat trace shorted. A passerby noted a small curl of smoke originating from the shorted wiring and also noted that the heat trace appeared to be glowing hot at one point. He reported the problem, and fire protocols were then initiated.

The subsequent investigation found that when the heat trace wiring had been initially installed in the early 1970s, the National Electric Code (NEC) did not require that ground fault protection, similar to the GFI (ground fault interrupted) devices now found in most bathroom outlets in homes built since the late 1970s, also be installed.

It was recommended by the manufacturer, however, that ground fault protection be installed because the vendor recognized at that time that the usual circuit protection of breakers or fuses would not protect against an internal short circuit within the heat trace wiring itself.

It was also found in the investigation that the vendor recommended that the heat trace wiring be periodically tested by meggering, a method for measuring levels of resistance and dielectric strength in insulating materials. This was recommended so that any age-related degradation or damage that might inadvertently occur to the internal dielectric materials used in the heat trace wiring to make heat could be detected and monitored.

As an aside, some time after this particular heat trace wiring had been installed the NEC did adopt a requirement to install protective ground fault sensors. Also, various other code publishers, such as the Institute of Electrical and Electronic Engineers (IEEE), adopted requirements for installation megger testing and also recommend periodic megger testing.

Thus, it was concluded in the investigation that the wiring likely was damaged by being kicked or stepped on by nearby work activities, and as a result of that there was sustained moisture infiltration into the dielectric material that caused the heat trace to short-circuit. It was further concluded that had the vendor recommendations been followed—installation of a ground fault device and periodic meggering to monitor the condition of the heat trace wiring—as soon as the short began to develop, either the ground fault device would have opened the circuit and provided an indication or the out-of-spec condition would have been noted. This would have then prevented the smoke and glowing condition, and the subsequent initiation of station fire protocols.

Note that the several time intervals are also depicted in Figure 4.10. In researching the problem, it was noted that heat trace wiring sometimes shorted and failed after about 10 years or so due to age-related effects, notably by moisture infiltration into the dielectric.

Investigation Strategies

Scientific Method

There is no reason why just one hypothesis at a time has to be considered. To efficiently use time, several hypotheses can be put forward, investigated in parallel to one another, and compared to the evidence as it is discovered. In this fashion, some hypotheses can be eliminated, and some modified in the iterative fashion already described in prior chapters. The arrangement noted in Table 4.4 for organizing data works reasonably well to do this.

Fault Tree Analysis (FTA)

This technique originated in the Bell Telephone Laboratories in the early 1960s to evaluate safety in the Minuteman Missile Launch Control System. As noted in Figures 4.5 through 4.8, fault tree analysis is a logical, cause-and-effect tree diagram method. The starting point is the failure, crime, or accident event, which is the final effect. The possible and most generalized immediate causes are then considered and either eliminated or supported by the evidence. This can be called level 1. Those causes not eliminated or falsified by the evidence then become the resulting effects for level 2, and the next, more detailed causes are considered. This continues, level by level, until the causes are known to sufficient detail commensurate with the investigation.

Sometimes, Boolean logic symbols are used, such as OR gates, EXCLUSIVE OR gates, AND gates, EXCLUSIVE AND gates, NOR gates, and so on, to indicate if–then type of cause and effect. The use of Boolean logic is useful in cases where an event could be caused by either single causes or multiple factors that must be present simultaneously to cause the effect. Sometimes, the probability of a particular cause-and-effect combination is known from experience, like that in Table 4.3. Using this, the branches of the diagram can be labeled with probabilities such that it becomes a reverse risk diagram.

Using Boolean symbols to lay out all possible failure modes beginning with the failure event and finishing with the potential causes creates an upside-down tree diagram. In some complex tree diagrams, if all the possible failure pathways are mapped, it can be seen that sometimes there are several possible pathways between the same initial cause and the final effect.

The specific path leading backwards in time from the failure event to the initiating cause or set of causes is sometimes referred to as the cut-set. When a full tree is flowcharted and there is more than one pathway from a specific cause, or set of causes, to the final failure, especially if programmatic processes and procedures are involved, the shortest realistic path between the two is called the minimum cut-set.

Using failure probability trees in this way can sometimes be a useful guide to determine where to focus investigative resources in a timely fashion. Some investigators pursue the most probable causes after initially doing sufficient fieldwork to eliminate, or falsify, sometimes on a conditional basis, the less probable ones to ensure they are not a factor.

As noted previously, fault trees are essentially reverse failure modes and effects analyses (FMEAs). Sometimes there are historical databases of failure probabilities available, or those probabilities are embedded in troubleshooting guides provided by the manufacturer, or by independent publications such as those from the Electric Power Research Institute (EPRI).

There are two software programs often cited for computerized fault tree analysis. The first is called CAFTA, which was developed by EPRI. Many power plants use this, especially in the nuclear industry, as well as some aerospace companies. The other software program is called SAPHIRE, which is used by some U.S. government regulatory agencies and bureaus, especially those dealing with atomic reactors and manned space vehicles.

The U.S. Nuclear Regulatory Commission's *Fault Tree Handbook*, NUREG-0492, published in January 1981, and the update published by NASA, *NASA Fault Tree Analysis with Aerospace Applications*, can be downloaded in PDF format free from the Internet, or paper copies can be purchased from the U.S. Government Printing Office for a nominal price. This publication is essentially an entire course in fault tree analysis theory and application.

Why Staircase

This method is not much different than the fault tree analysis method or the event and causal analysis method. Some people call it "the little kid" method of investigation, since the method simply consists of a series of *why*'s being asked.

It starts out with defining what happened, that is, quantifying the resulting undesired effect. It then asks what the immediate cause of the resulting event was. When this first-level *why* is answered, and it should also answer the question "Why is this the immediate cause and not something else?" the next level *why* is then queried. This iterative continues until a sufficient *why* level is reached at which reasonable control can be exercised over the causes. Usually, an apparent cause type report needs to go only two or three *why*'s deep. A more comprehensive root cause investigation may go four, five, or even six *why*'s deep.

For example, a resulting undesirable consequence may be the end result of a chain of events that all emanate from a single decision made by a subcommittee that was approved by the whole committee. At this point, another

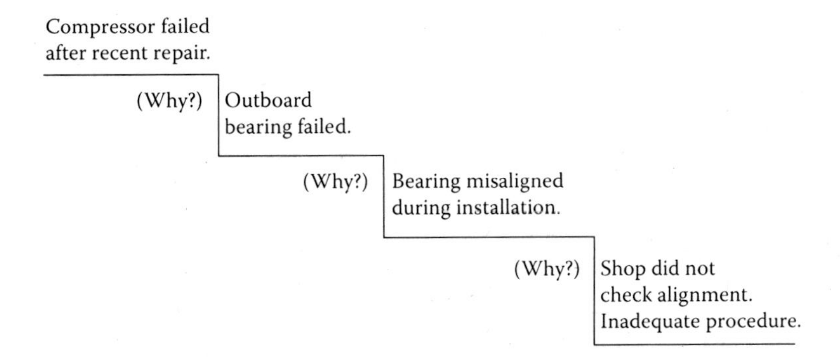

Figure 4.11 *Why* staircase example of a compressor failure.

why level could be instituted to investigate the psychological profile of each of the subcommittee and committee members to assess their motivations for making that decision. However, it might be more effective from the standpoint of time, resources, and managerial control to simply institute some decision-making guidelines on the subcommittee, such as an independent expert panel review.

The *why* staircase method is often depicted graphically as a series of descending steps. The top step is the final, undesirable consequence. Each step below the top step is the immediate, preceding cause. It looks something like Figure 4.11.

Human Performance Evaluation Process (HPEP)

Human error plays a role in many accidents, failures, crimes, and other types of undesirable consequences. The term *error* is used in this context to specially refer to actions taken by a person that were not intentionally malicious or destructive.

A human error can directly cause an undesirable event; it may make the event worse, more severe in scope, or cause it to last longer; or it may have helped foster the event by creating the conditions that increase the probability that the event would happen. Human error includes simple mistakes such as pushing the wrong button, inappropriate responses to conditions and circumstances such as not calling the fire department before attempting to put the fire out, or perhaps inadequate responses to conditions and circumstances such as not examining the rest of the herd when one cow is suspected of having mad cow disease (bovine spongiform encephalopathy).

When human error has played a significant role in the event, it is useful to determine whether the error was random or if there were specific conditions that increased the probability that the error would be made. For example, road fatigue and driver distractions play a significant role in the causation of car accidents.

Some general categories of factors that affect human performance include the following:

Fitness. This includes a person's general health, a person's mental attitude, and the influence of drugs or medications, fatigue, and physical or mental stress.

Knowledge. This includes the knowledge and training to do the task, skill level, proficiency level, cognitive abilities and attentiveness with respect to the task, familiarity with the task, and so on.

Procedures and instructions. Are they sufficiently detailed? Do they read well? Are they confusing? Are they conditions not covered in the procedure and are ambiguous? Do they contain error traps, that is, decision points within the procedure where it is easy to make a mistake?

Tools. Did the person have the right tools? Was he having to make do? Was the person wearing the appropriate protective clothing and equipment? Was all the safety equipment functional as intended?

Supervision and staffing. Were enough people assigned to the task? Was there adequate oversight by experienced personnel? Were decisions made promptly at the correct level? Did everyone know who was in charge, or was there confusion about who was to make the decision?

Ergonomics. Was the machine and human interface deficient? Were there appropriate warning alarms? Was the machine hard to use? Was the information displayed in an easy-to-understand format? Did the equipment respond as expected? Did the operation of the equipment require some kind of special knack that only some people understood?

Environment. This includes factors that can affect a person's ability to work, think, understand, function, and communicate, such as sound and noise levels, lighting, temperature, humidity, visual distractions, chair design, clothes, shoes, colors, bugs, and smell, to name a few.

One popular and simple-to-use flowchart for depicting the factors involved with deficient human performance was developed for use by the Nuclear Regulator Commission and is contained in NUREG/CR-6751, *The Human Performance Evaluation Process*, which was published in 2001 and is listed in the reading list. The flowchart is reproduced in Figure 4.12. The numbers in the boxes in the far right of the figure refer to chapters in the monograph from which this was taken, where the topic is discussed in greater detail (see reading list).

Change Analysis

Change analysis looks back to when a task or activity has been previously successfully accomplished, or perhaps uses an ideal model of how the task

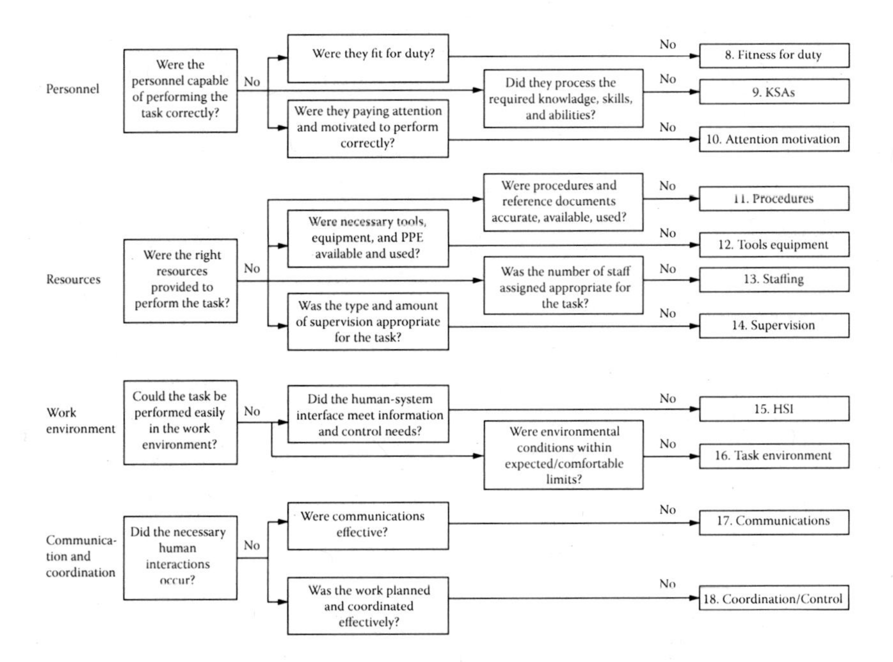

Figure 4.12 Human performance evaluation process cause tree. (From NUREG/CR-6751, September 2001.)

or activity should be accomplished. A comparison is then made to when the task or activity was not done successfully to determine what has changed or what is different between the two results. The differences between when the task or activity was successful and when it was unsuccessful are often called the deltas or gaps.

One way to graphically depict a change analysis is to flowchart the successfully completed task or event and then similarly flowchart the unsuccessful event alongside it. This will highlight the deltas. The differences, or deltas, must then be individually assessed to determine whether they logically caused or contributed to the failure. Not all the deltas, of course, may be relevant to having caused or contributed to the failure. (See the "Coincidence, Correlation, and Causation" section in Chapter 3 where *post hoc ergo propter hoc* is disussed.)

Barrier Analysis

Barrier analysis is often used when barriers have been set up to prevent dangerous, unsafe, or hazardous actions from occurring. The barriers may be physical devices, such as fuses, breakers, fences, guards, or locked cabinets. They may also be administrative barriers, such as rules, regulations, laws, company policies, training certifications, technical codes, technical specifications, design margins, or redundant design features.

The idea behind barrier analysis is that because barriers have been previously put in the way to prevent the undesirable event from occurring, a barrier must either have failed, was not in place, or was somehow circumvented for the failure, accident, or crime to have occurred. In a crime, of course, the circumvention was likely deliberate and considered.

One way to graphically present a barrier analysis is to flowchart the event with a basic timeline. Then add into the timeline the places where the barriers should have prevented the undesirable event from going further, causing the failure.

Subjectively, some barriers are considered strong and some weak. A physical device that prevents an unsafe act from occurring at all is usually considered to be a strong barrier. It is judged to be effective in preventing or stopping the undesired event from occurring. A safety guard over a flywheel is an example of a strong barrier. If such a guard were removed for maintenance and was not replaced when the maintenance was completed, the loss of this barrier could cause an accident.

An example of a weak barrier is training that was done a long time ago and was done only once. Because training is often forgotten when not reinforced or practiced regularly, its effectiveness diminishes with time. Even if a person was once well trained about how to do a complex task, if he has not done that task in a long time, he has an increased risk of making a mistake.

Some barriers are explicit and some are implicit. An example of an explicit barrier is a safety rule that is posted, enforcing the wearing of a hard hat. An implicit barrier might be personal shame that would be experienced if a person were caught doing something considered immoral or dishonest, or perhaps peer disapproval of certain actions.

In some cases, it has been necessary to overcome the implicit barrier of peer disapproval to put into place explicit safety rules. An example of that is the use of earplugs. At some locations, despite the obvious ear damage that can result from not using them, some male workers have in the past considered earplugs to not be manly. This collective peer pressure sometimes caused other male workers to also not use earplugs, so that they, too, would be considered macho and avoid ridicule by the group.

To increase the odds that there will be no failure or undesired outcomes, it is typical that there are several layers of barriers. It is presumed that each successive barrier reduces the chances that the failure will occur. Consequently, when failures do occur, it is sometimes the case that there are a series of things that went wrong all at the same time.

The Space Shuttle *Challenger* accident that occurred January 28, 1986, and the Space Shuttle *Columbia* accident that occurred February 1, 2003, are two famous events often cited as fitting this category. Both failures were the result of a series of flawed decisions, poor judgments, and equipment deficiencies that had been considered barriers to prevent such accidents from occurring.

In that regard, it is worth repeating verbatim two paragraphs contained within the executive summary in Volume 1 of the Columbia Accident Investigation Board report of August 2003. They describe how multiple layers of barriers erected by even the brightest and best can fail and allow an event such as the *Columbia* accident to occur.

> The physical cause of the loss of Columbia and its crew was a breach in the Thermal Protection System on the leading edge of the left wing, caused by a piece of insulating foam which separated from the left bipod ramp section of the External Tank at 81.7 seconds after launch, and struck the wing in the vicinity of the lower half of Reinforced Carbon-Carbon panel number 8. During re-entry this breach in the Thermal Protection System allowed superheated air to penetrate through the leading edge insulation and progressively melt the aluminum structure of the left wing, resulting in a weakening of the structure until increasing aerodynamic forces caused loss of control, failure of the wing, and break-up of the Orbiter. This breakup occurred in a flight regime in which, given the current design of the Orbiter, there was no possibility for the crew to survive.
>
> The organizational causes of this accident are rooted in the Space Shuttle Program's history and culture, including the original compromises that were required to gain approval for the Shuttle, subsequent years of resource constraints, fluctuating priorities, schedule pressures, mischaracterization of the Shuttle as operational rather than developmental, and lack of an agreed national vision for human space flight. Cultural traits and organizational practices detrimental to safety were allowed to develop, including: reliance on past success as a substitute for sound engineering practices (such as testing to understand why systems were not performing in accordance with requirements); organizational barriers that prevented effective communication of critical safety information and stifled professional differences of opinion; lack of integrated management across program elements; and the evolution of an informal chain of command and decision-making processes that operated outside the organization's rules.

In this regard, I highly recommend that Volume 1 of the Columbia Accident Investigation Board report of August 2003 be read, especially the executive summary, in its entirety. It is a free download from the Internet. The Internet web address and other publication particulars are listed in Chapter 8.

Motive, Means, and Opportunity (MMO)

In a previous section that discussed the human performance evaluation process (HPEP), it was tacitly implied that the failure to act appropriately was perhaps inadvertent, accidental, or unintentional. Perhaps if the person had been better informed, better equipped, more rested, better trained, or had better instructions or better leadership, the mistake might have been avoided. In other words, it was implied that the person wanted to perform appropriately but was unable to do so for some reason.

There is, of course, a dark side of human performance that involves deliberate actions or inactions that result in harm, damage, or loss. When the actions of a person that initiate a failure, accident, or crime appear to have been deliberate, it is useful to assess means, motive, and opportunity.

Means is defined as the ability of a person to do the action necessary to cause the result. Does he have the skill, the dexterity, the resources, the equipment, or the physical and mental attributes to cause the result?

Motive is defined as the reason why a person would do the act necessary to initiate or allow harm, damage, or loss to occur. Since there is often either an implicit or explicit social, legal, or administrative barrier to being caught doing the act, a sufficient motive is necessary to overcome the fear or reticence associated with deliberately violating a known barrier.

It has been reported by various sources that there are seven basic motives for a crime or crime-like action. Whether this list is completely exhaustive is open to question, but it does appear to encompass most of the motives an investigator would regularly encounter. They are as follows:

1. Gain: This is one of the most popular motives. Gain, which is usually money or property, is either for the person or for a favored second person, like a relative or romantic partner.
2. Anger, revenge, vengeance.
3. Covering up another crime or action that the first person wishes to remain hidden.
4. Jealousy.
5. Excitement and thrills: This is often associated with abnormal mental conditions such as those induced by drugs, psychoses, neuroses, irrational impulses, perversions, temporary mental states, unusual physical condition, and so on. Pyromaniacs fit into this category. Certain usual types of sexual gratification may also fit into this category.
6. Vandalism, civil riots, revolutions, insurrections, and war.
7. Terrorism: Fanatical religious or political beliefs that require hostile or resistive actions as part of their belief system.

Opportunity is simply whether a person was in the right place at the right time to accomplish the act or deed.

Table 4.5 MMO Summary Table

	Suspect A	Suspect B	Suspect C
Motive	X	X	?
Means		X	X
Opportunity	X	X	X
Physical link		X	

It is important to recognize that just because a person has the motive, means, and opportunity, does not mean that the person actually has done the deed. It is possible that several people may have the same motive, means, and opportunity. In the case of sabotage on the job, for example, this is often the case when there are hard feelings between labor and management.

Thus, having all of these attributes does not, by itself, constitute proof. It might be said that means, motive, and opportunity are necessary, but not sufficient. This is why it is important to assiduously apply the principle of falsification.

Motive can be further subdivided into weak or strong motives. Sometimes it is difficult to detect the real motive of a person. For various reasons, the culprit may not even know himself if the deed was done on impulse. However, Cicero, the famous Roman litigator, statesman, and orator said it best when he posed the question: *Cui bono?* (Who benefits?)

Timelines and verifiable alibis are useful in determining whether a person had the opportunity to commit the act, deed, or crime. Records and document checks are useful in determining whether a person had the means. For example, if a certain tool was needed to commit the act, determining who has such a tool, where such tools might be found, and who has and does not have access to them is very useful information.

If there is a list of suspects, motive, means, and opportunity information can be handily organized, for example, as in Table 4.5.

Note that in Table 4.5, a fourth item was added to the list: physical link. Finding a unique physical link between a suspect and the crime or deed further increases the likelihood that the particular suspect may have been the perpetrator.

Management Organizational Risk Tree (MORT)

This method was developed in the early 1970s for the U.S. Energy Research and Development Administration to assess safety-related issues in complex, goal-oriented organizations. There is also a simpler variant of MORT, called SMORT (Safety Management Organizational Review Technique). Essentially, it is an organizational fault tree analysis method.

In this method, the various safety programs and program elements are flowcharted logically. The undesired consequence, which often is poor performance, a failure, or an accident, is the starting point. The deficiencies that caused the undesirable consequence are tracked through the various programs or program elements in the organization that had control over them.

MORT tracks who or which program was responsible for this, who or which program was responsible for that, when was this supposed to happen, and so on.

In this way, all the programmatic deficiencies in the organization that led to or contributed to the undesirable consequence are identified. Once identified, the reasons why each program or program element failed to do its part can be individually assessed and fixed.

Proprietary Investigation Methods

There are many popular proprietary investigation methodologies currently being packaged and sold. There are at last count over one hundred. Many, if not most, proprietary investigation methods come with training manuals, both online and offline, that are replete with their own specialized jargon. Of course, they also come with professional trainers to teach the method at your company's location, or at a school where people attend a 1-, 3-, or 5-day how-to course of instruction.

Most proprietary investigation methodologies include a supply of copyrighted formatted flowcharts and worksheets for interim reports and final reports, and a range of publications that help the novice step through the particular methodology. Some also offer linked computer flowcharting and word processing programs that produce complete reports formatted according to the particular methodology. Besides the usual intensive initial training session for novitiates, there are also some advanced training classes where a person can become a certified trainer in a particular method.

Among the more popular proprietary investigative methodologies being marketed are Proact®, Kepner–Tregoe® (i.e., eThink® and Project Logic®), Apollo® (i.e., RealityCharting®), and Tap Root®. Other failure or root cause training materials are known by the name of the primary instructor who initially developed the materials, e.g., Chong Chiu of FPI International®. However, this is just a small, token list. There are too many to mention by name.

Possibly the most well-known and well-marketed proprietary method is the Kepner–Tregoe method, commonly called the K–T method, whose company headquarters is in Princeton, New Jersey. The founders of the method, Dr. Kepner, a psychologist, and Dr. Tregoe, a sociologist, discovered or concluded in the 1950s, while employees at the Rand Corporation, that good decision making in the Air Force had more to do with exercising a logical process in gathering, organizing, and analyzing information than it did with an individual's rank. (Please read the *Columbia* Accident Investigation Board report executive summary with this in mind.) This finding became the basis for the "rational process" espoused in their various business analysis methods. The rational process method was also explained in the 1965 book *The Rational Manager*, which was published by the McGraw-Hill Company. Most of the K-T method is based upon the principles discussed in the book.

In brief, the rational process is a structure that purports to maximize critical thinking skills of key people. The rational process includes the following fundamental parts:

- A situation appraisal
- Problem analysis
- Decision analysis
- Potential problem analysis

The remarkable success achieved in marketing the K-T method in its various forms, to both domestic and international businesses, has, not surprisingly, spawned many competitors. Consequently, most organizations of any size eventually will be approached by representatives of some of these proprietary investigative method companies, or at least receive materials from them in the mail that urge the management of the organization to purchase their training materials, lessons, or programs.

No one proprietary method appears best. Each has its good points and weak points, depending upon the particular application, the instructor, and the ability of the person to apply the particular method. Some seem to work better for management problems, and some seem to work better for equipment problems.

Good proprietary investigation methods can be helpful in organizing and managing an investigation. Engaging the services of one of these providers also relieves an organization from the burden of preparing its own training materials, especially if the organization internally lacks the experience, time, or resources to do so. Graduates of the various proprietary methods then often become the investigation managers or investigation team leaders within their company or organization.

Knowing a particular proprietary method, however, does not automatically imbue graduates with Sherlockian acumen and is not a substitute for subject matter knowledge. Sometimes, newly minted graduates of a particular proprietary method become enamored with the method and promote its use for all occasions and circumstances with the zeal of a missionary. In such cases, I am reminded of the quip, "If you happen to be a hammer, the rest of the world looks like a nail."

Those proprietary investigation methods that eschew the scientific method, utilize implicit assumptions in their analytical techniques, are not evidence driven, do not pursue falsification, and do not readily allow initial hypotheses to evolve as evidence is collected should be avoided. Without these fundamentals, the resulting investigation report may simply be a collection of logically disjointed facts and opinions, slickly packaged in a nice-looking format.

Four Common Reasons Why Some Investigations Fail*

5

The evil that men do lives after them;
the good is oft interred with their bones.

William Shakespeare, from *Julius Caesar*, 1599

Introduction

Figure 5.1 shows an x-ray radiograph of a 1N752A type Zener diode taken by Stan Silvus, manager of the Component Analysis Laboratory at Southwest Research Institute. The radiograph shows that this particular Zener diode was not manufactured with die-attach at one end of the die and had only marginal die-attach at the other end. Consequently, the die-attach deficiency caused the Zener diode to fail unexpectedly in an intermittent fashion. In turn, this caused a failure in the voltage regulator system of an emergency diesel generator system, which then caused the emergency diesel generator to be temporarily taken out of service.

The failure of this particular Zener diode occurred in a circuit board that had less than 40 hours of actual service time, although the circuit board itself was over 27 years old. It had been a spare board kept in inventory.

Going to this level of detail to gather evidence may seem extreme to some. However, this particular evidence was fundamental to validating the hypothesis that the failure event cause was a random failure due to a manufacturing defect, and falsifying the hypothesis that the failure was caused by an infant mortality type failure. In the nuclear power industry, this distinction is significant.

Reason 1: The Tail Wagging the Dog

As the investigation proceeds and information about the failure event accumulates, some initial cause hypotheses can be readily falsified by the

* A condensed version of this chapter was originally published in the December 2007 issue of *Maintenance Technology* under the title "Why Some Root-Cause Investigations Don't Prevent Recurrence." Applied Technology Publications has courteously permitted the republication of material, especially the introduction and the first three reasons, from that article for this chapter.

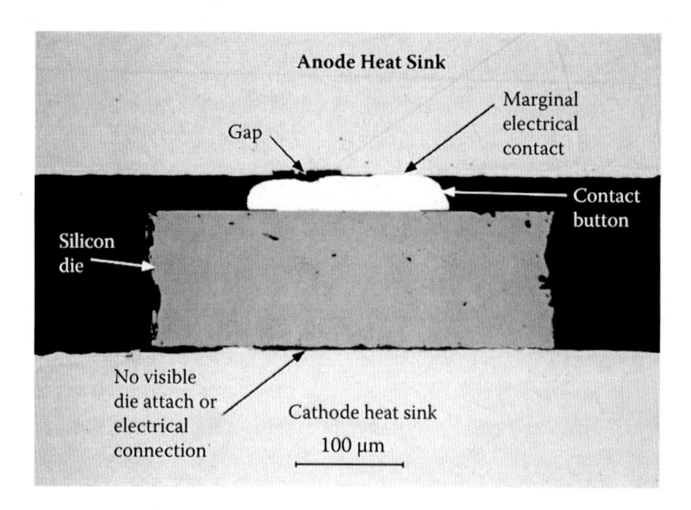

Figure 5.1 Cross-sectional view of a Zener diode that caused a failure in a nuclear station emergency diesel generator system.

preliminary evidence and dismissed from further consideration. The smaller group of remaining hypotheses will likely share some common characteristics. Consequently, these remaining hypotheses will likely require more effort to determine which ones specifically apply.

At this point in the investigation it may become apparent what the final conclusion concerning the basic cause of the event might be. Based upon the evidence gathered so far, perhaps only two or three viable cause hypotheses remain in consideration. Consequently, it also becomes apparent what the corresponding corrective actions or follow-up actions might be for those causes.

By anticipating which corrective or follow-up actions are more palatable to the client, to management, or to the governing authorities, the investigator may begin to subconsciously, or perhaps even consciously, steer the remainder of the investigation to arrive at a conclusion whose corresponding follow-up actions are less troublesome. Evidence that appears to support a cause that leads to more palatable follow-up actions is actively sought, while evidence that might falsify the favored cause is not actively sought.

Evidence that might falsify the favored cause may be dismissed as being irrelevant or not needed. It may be tacitly assumed to not exist, to have disappeared, or to be too hard or expensive to find. It may even be simply ignored because so much evidence already exists to support the favored cause that the investigator presumes he already has the answer.

As previously noted, this is *a priori* methodology. Essentially, an outcome or conclusion is decided before the investigation has been completed. In this case, the investigator has presumed what follow-up or corrective actions are reasonable to do. Subsequently, he uses the remainder of the investigation to seek evidence that points to a cause that corresponds to the palatable follow-up actions he desires.

Here is an example of this. A close-call accident involved overturning a large, heavy, lead-lined box mounted on a relatively tall, small-wheeled cart. The failure event investigation team found that the box and wheeled cart combination was intrinsically unstable. The top-heavy cart easily tipped when the cart was moved and the front wheels had to swivel, or when the cart was rolled over a carpet edge or floor expansion joint.

The investigation team also found that the personnel who had moved the cart in the course of doing cleaning work in the area did so in violation of an obviously posted sign. The sign said that prior to moving the cart a supervisor was to be contacted. However, the personnel inadvertently moved the cart without contacting a supervisor in order to clean under and around the cart.

The easy corrective actions in this case are to chastise the personnel for not following the posted rules and to strengthen work rule adherence through training and administrative permissions. There is ample evidence to back-fit a failure event cause to support these actions. Such a cause finding, and its corresponding corrective actions, is consistent with what everyone else in the industry has done to address the problem as noted in ample operational experience reports.

Further, this course of action is attractive because it appeals to notions of personal accountability, is cheap to do, and can quickly dispose of the problem. In short, the close-call accident was blamed on the workers for not following the rules. The inherent instability problems found to exist in the box-and-cart combination were ignored.

Unfortunately, when the cart-and-box combination is rolled to a new location, the same problem may recur because the cart-and-box combination is still unstable. The new administrative permissions and additional training may reduce the probability of recurrence, but it will not eliminate it. When the cart is rolled many times to new locations, it is probable that the problem will eventually recur and perhaps cause a significant injury. It is similar to the hockey analogy of "shots on goal." Even the best goalkeeper can be scored upon if there are enough shots on goal.

Reason 2: Lipstick on a Corpse

In this instance, the failure event has been successfully investigated. An event cause has been concluded that is supported by ample evidence. Vigorous attempts to falsify the cause conclusion have failed. OK, so far, so good.

However, perhaps the cause conclusion involves a deficiency involving a friend of the investigator, a manager known to be vindictive and sensitive to criticism, or some company entity that cannot bear criticism due to previous problems. The latter could include an individual who might get fired if he is found to have caused the problem, an organization that might be fined or sued for violating a regulation or law, or a department that might be reorganized or eliminated for repeatedly causing problems.

In other words, the investigator is aware that the actual consequences of identifying and documenting the cause of the event will be greater than just the corrective actions themselves.

When faced with this dilemma, some investigators attempt to wordsmith the failure event cause report in an effort to minimize perceived negative findings and emphasize perceived positive findings. Instead of using plain, factually descriptive language to describe what occurred, less precise and more positive-sounding language is used.

Wordsmithed reports are relatively easy to spot. Instead of using plain modifiers such as *deficient* or *inadequate* to describe a process, phrases such as "less than sufficient" or "less than adequate" are used. Instead of reporting that a component has failed a surveillance test, the component is reported to have had "less than the target output" or perhaps "met 95% of its expected goals."

In such cases, the event cause report becomes a quasi-public relations document that sometimes has conflicting purposes. Because it is an event cause report, its primary purpose is supposed to be a no-nonsense, fact-based report that documents what went wrong and indicates how to fix it. However, a secondary purpose is introduced, which is to convince the reader that the failure event and its underlying cause are not nearly as bad as the reader might otherwise think.

With respect to recurrence, there are two fundamental problems with wordsmithing a failure event cause report. The first is that it blurs or mitigates the effectiveness of the corrective or follow-up actions. Corrective actions work best when they are specific and targeted. A diluted or minimized cause, however, is often matched to a diluted or minimized corrective action.

There is a strong analogy to the practice of medicine in this instance. When a person has an infection, if the degree of infection is underestimated, the medicine dose may be insufficient and the infection will recur. This is also true for failure event investigations.

The second problem is that by putting a good "spin" on the problem, management or the authorities may not properly support what needs to be done to fix the problem. In other words, the report succeeds in convincing its audience that the failure event is not a serious problem that warrants serious attention.

Reason 3: Elementary, My Dear Watson, Elementary

In some ways, failure event investigations are a lot like whodunit novels. Some plant or office personnel just cannot resist making a guess about what caused the failure in the same way that mystery buffs often try to second-guess who will be revealed to be the murderer at the end of the story. It certainly is fun for a person, and perhaps even a point of pride, if his initial guess turns out to be right. However, there are circumstances when such a guess can jeopardize the integrity of a failure event investigation.

The circumstances are as follows:

- The guess is made by a senior manager involved in the investigation process.
- The plant or facility has an authoritarian, chain-of-command style organization.
- The management culture puts a high premium on being right and has a zero-defects attitude about being wrong.

The scenario goes something like this:

- A failure event occurs or a condition adverse to quality is discovered.
- Some preliminary data are quickly gathered about conditions in the plant when the failure occurred.
- From this preliminary data, a senior manager guesses that the fundamental cause will likely be x, because
  1. he was once at a plant where the same thing occurred or has read about it in an article; or
  2. applying his own engineering acumen, he deduces the nature of the failure from the preliminary data like a Sherlock Holmes or a Miss Marple.
- Not being particularly eager to prove their senior manager wrong and deal with the consequences, the investigation team looks for information that supports his hypothesis.
- Not surprisingly, they find some; the presumption is then made that the cause has been found and fieldwork ceases.
- A report is prepared, submitted, and approved, possibly by the same senior manager who made the Sherlockian guess.
- The senior manager takes a bow, once again proving why he is a senior manager.

The deficiency in this scenario that can lead to failure event recurrence is the fact that falsification of the favored hypothesis was not pursued. Once a cause was presumed to have been found, significant evidence gathering ceased. (Why waste resources when we already have the answer?) Consequently, evidence that may have falsified the hypothesis, or perhaps supported an alternate hypothesis, was left in the field.

Again, this is another example of an *a priori* methodology: where the *de facto* purpose of the investigation is to gather information that supports the favored hypothesis. In short, the fundamental cause of the event was missed, a seemingly plausible one was accepted, and the fundamental problem that caused the event in the first place was not fixed or addressed.

In this regard, there was a famous experiment conducted by Daniel Simmons of the University of Illinois and Christopher Chabris of Harvard University about directed observation that applies. Test subjects in the experiment were told to watch a type of basketball game carefully because they would be questioned about how many times the basketballs would be tipped into the air by the participants.

Two teams of three players on each tossed around two basketballs for 1 minute in a video. One team was dressed in white shirts, the other in black shirts. Test subjects were asked to count the number of passes made by the white team during the "game." After 35 seconds, a person in a gorilla suit walked into the room, beat his chest, and 9 seconds later, exited. Incredibly, only 50% of the viewers tasked with counting passes noticed the gorilla. They were focused on counting passes.

When the test subjects who did not see the gorilla were told about it, they were incredulous and did not believe that they had missed seeing someone masquerading as a gorilla—until they were shown the tape a second time. At that point, they all observed the gorilla. This effect, of not seeing the obvious because of being focused on a specific task or activity, is sometimes called inattentional blindness.

Reason 4: Dilution of the Solution

The following situation is usually peculiar to large organizations with many administrative processes that have individual departments whose departmental priorities are not in alignment with company priorities.

The investigation of a failure condition has been completed and was well done. The root cause of the condition and all contributing causes were properly identified. An excellent corrective action plan was crafted to fix the problem. The plan was presented to senior management. Senior management approved the plan, wholly supported the plan, and provided adequate resources to execute the plan. The problem is fixed, right?

Not so fast. Let's say that the original failure condition involved a deficient quality control process in the company's safety program. Specifically, too many injuries and accidents had occurred in the plant in the past 2 years. Consequently, insurance costs had increased, the number of lost workdays had increased, and regulatory scrutiny was being threatened due to the resulting high OSHA reportables rate.

Most quality control processes involve four parts. For a safety program, the four parts are as follows:

- Establishment of specific safety standards that are to be followed
- Communication of those standards to the people who have to follow them
- Reinforcement of those standards by reminders, mentoring, training, and supervisory oversight
- Enforcement of those standards

In the investigation, it was found that only the enforcement part was being done, and even that was not being done consistently. Different divisions of the company had different safety practices. No training of employees and supervisors about company safety rules had been done in years despite an influx of new, inexperienced employees. Supervisors and managers did not remind employees prior to a job being undertaken what safety rules needed to be observed and what safety gear had to be worn. When some supervisors noted that safety gear was not being utilized, because they were unsure of what company policy was, they often did not say anything to the employee.

The investigation noted that unless all four parts of the process were done effectively and in order, low accident and injury rates could not be achieved. In reviewing both successful and unsuccessful safety programs at similar plants, when all four parts were executed well and in an orderly fashion, injury and accident rates were low. In plants where one, two, or more of the parts were executed poorly or not even performed, accident and injury rates were higher.

In investigating the situation, the root cause investigation team had used a modified form of the barrier method to figure out what was wrong and what should be done. First, a model program that worked well was flowcharted. Then, the actual program used at the company was flowcharted the same way so that it could be compared to the model program, item by item. Where there was a difference between the model and the company's program, a corrective action was proposed to fix the discrepancy.

However, when it came time to actually execute the various corrective actions, the plan fell apart. The manager tasked with developing a consistent, company-wide set of safety standards and practices had other, higher-priority tasks to do and had vacation already scheduled. Consequently, he had the scheduled due date extended to fit his personal and departmental schedule. The extension was easily approved since that manager had considerable administrative clout, and the approving authority was unaware of the importance of this task being done on time with respect to the overall corrective action plan.

With no consistent safety standard and practices in place, the people responsible for training employees did not know what training to provide.

However, because they did not want to miss their scheduled deadline, they trained personnel in accordance with the patchwork of safety standards that already existed. This further institutionalized the existing confusion about safety practices.

Since the training for supervisors did not include the new material about how to reinforce safety practices by performing prejob briefs and supervisory coaching techniques, there seemed to be no difference in the new company safety policy from the old one. Consequently, the reinforcement part of the plan lapsed.

Lastly, because the training was confusing and inconsistent, supervisors were not sure what practices they should enforce. Thus, enforcement of existing safety practices continued more or less as it had been done before.

The end result is that the well-considered corrective action plan did not readily fix the problem as intended. While no one was apparently looking, a company goal, the establishment of a consistent company safety program in accordance with good quality control process practices to lower accident and injury rates, was subverted to oblige a departmental scheduling convenience.

This above example brings to mind a quotation attributed to Benjamin Franklin:

> For want of a nail the shoe was lost.
> For want of a shoe the horse was lost.
> For want of a horse the rider was lost.
> For want of a rider the battle was lost.
> For want of a battle the kingdom was lost.
> And all for the want of a horseshoe nail.

In the preceding example, the fundamental problem was that the corrective action plan suffered from exactly the same fundamental problem that had been identified by the root cause investigation. The same managers and departments were put in charge of fixing the problem that had created in the first place, they approached the work with the same attitude concerning its importance, and they ended up with the same result.

To successfully circumvent this pitfall, the corrective action plan schedule should have been delegated to an authority one level above the departments involved in the execution of its various parts. This person should have had the authority to muster resources and establish priorities to ensure that the parts of the plan were completed on time and coordinated properly, in the manner of a project manager.

In short, a plant-level goal should be managed by a plant-level manager. Putting departmental managers in charge of a plant-level goal subverts the plant-level goal into a departmental project.

Report Writing
Three Case Studies

6

Management is doing things right; leadership is doing the right things.

Peter Drucker, 1909–2005,
developer of the first MBA program in the United States

Reporting the Findings of an Investigation

A well-written report that explains plainly what happened, how it happened, and why it happened is valuable and serves many purposes.

There are several formats used to report the results of an investigation. The easiest format is a simple narrative, where the investigator simply describes all his investigative endeavors in chronological order. He starts from when he received the telephone call from the client and continues until the last item in the investigation has been completed. The report can be composed daily or piece-wise when something important occurs as the investigation progresses, like a diary or journal. Insurance adjusters, fire investigators, and detectives often keep such a chronological journal to record their findings. A narrative report is simply a more readable, edited version of such a journal or diary.

A narrative report works well when the investigation involves only a few matters and the evidence is straightforward. However, it can be difficult for a reader to imagine the event reconstruction from a simple narrative report when there are many facts to consider, when the line of reasoning is involved, when there are test results and laboratory reports that require evaluation, when there are eyewitness accounts that have to be cross-correlated to each other and to the laboratory reports, and when several interconnected scientific principles have to be understood and applied. The links and significance among these various items, which may be discovered at various times during the investigation, may not be readily apparent to the reader in a simple narrative. The investigation process itself may not proceed in a logical, step-by-step fashion. Also, sometimes answers show up before the questions are even fully developed.

Alternately, an investigative report can be prepared like an academic paper, replete with technical jargon, equations, graphs, and copious reference footnotes. This type of report, while appropriate for reporting research or the findings of a particular investigation to similarly experienced professionals, is often unsatisfactory. It may not readily convey the findings and assessments of the investigation to the variety of people who need to ultimately read, understand, and use the report.

To determine what kind of format to use, it is often best to first consider who will be reading the report. In general, the audience who might read a root cause investigation report includes the following:

- *Claims Adjusters*: The adjuster will use the report to determine whether a claim should be paid under the terms and conditions of the insurance policy. If he suspects there is subrogation potential, he will forward the report to the company's attorney for evaluation. In some insurance companies, such reports are automatically evaluated for subrogation potential. Subrogation is a type of lawsuit filed by an insurance company to get back the money they paid out on a claim by suing a third party that might have had something to do with causing the loss.

  For example, if a windstorm blows the roof off of a house, the insurance company will pay the claim to the homeowner, but may then sue the original contractor because the roof was supposed to withstand such storms without being damaged.
- *Law Enforcement Authorities*: They will certainly be interested if the report is related to a crime, criminal negligence, or a violation of the law. If that is the case, the report may form the basis for further investigation and may become part of their body of evidence if prosecution is pursued. They may compare the facts and findings in the report against what they may have already collected, and may repeat or otherwise verify any laboratory work to confirm findings and observations discussed in the report. If any evidence was collected and stored during the preparation of the report, law enforcement authorities may take possession of it.

  If there is apparent evidence or information in the report that indicates that there has been a crime, in most cases there is a professional and legal obligation to bring this to the attention of the proper authorities. The only exception that might apply to this obligation is when the principle of client–attorney privilege has been invoked. If this is the case, carefully clarify what your obligations are with the attorney of record in this matter.
- *Attorneys*: This includes attorneys for both the plaintiff and the defendant. The attorneys will scrutinize every line and every word used in the report. Often, they will inculcate meaning into a word or phrase that the investigator never intended. Sometimes the investigator will

inadvertently use a word in an engineering context that also has a specific meaning in a legal context. The legal meaning may be different from the engineering meaning. Lawyers are by trade wordsmiths. Engineers as a group are renowned for being poor writers. This disparity in language skill often provides the attorneys for either side with plenty of sport in reinterpreting the engineer's report to mean what they need it to mean.

- *Technical Experts*: The report will also be read by various technical experts who possess similar skills and knowledge as the members of the investigating team. They may be only professionally interested, or they may be auditing the report searching for technical errors or omissions. They will want to know upon what facts and observations the investigator relied, which regulations and standards he consulted and applied, and what scientific principles or methodologies were used to reach the conclusions about the cause of the loss or failure. The experts for the other side, of course, will challenge each and every facet of the report that is detrimental to their client and will attempt to prove that the report is a worthless sham. Whatever standard the investigator used in his report will, of course, be shown to be incorrect, incorrectly applied, or not as good as the one used by the other side's technical expert.

  One common technique that is used to discredit a report is to segment the report into minute component parts, none of which, when examined individually, are detrimental to their side. This technique is designed to disconnect the interrelationships of the various components and destroy the overall meaning and context. It is akin to examining individual liver cells, or perhaps the spleen in a person's body, to determine if the person is in love or roots for the Cardinals.
- *The Author*: Several years after the report has been turned in to the client and the matter has been completely forgotten about, the investigator who originally authored the report may have to deal with it again. Court cases can routinely take several years for anything to happen. Thus, several years after the original investigation, the investigator may be called upon to testify in deposition or court about his findings, methodologies, and analytical processes. Since so much has happened in the meantime, he may have to rely upon his own report to remind himself of the particulars of the case and what he did.
- *Judge and Jury*: If the matter does end up in trial, the judge will decide if the report can be admitted into evidence, which means that the jury will be allowed to read it. Because this is done in a closed jury room, the report must be understandable and convey the author's reasoning and conclusions solely within the four corners of each

page. It is well to bear in mind that the average jury has an educational level of less than 12th grade. Most jurors are uncomfortable with equations and statistical data. Some jurors may believe "there is something to" astrology and alien visitations, will be distrustful of "pointy-headed professors" from out of town, and since high school, their main source of new scientific knowledge has consisted of television shows and popular publications such as *The National Enquirer.*

- *Managers*: If the matter involved a process, machine, or event that occurred in a business setting, the people who manage the company will want to understand what is in the report and use the report as a basis for corrective action to prevent or mitigate something similar from occurring again. In some cases, a person might be fired based upon the findings of such a report; in other cases, it might form the basis for a significant investment in equipment or training.
- *Other Professionals*: Professionals in related industries may wish to avoid having this type of event occur at their plant or facility, or perhaps it has already occurred and they wish to see if what they experienced has factors in common with what is described in the report.

In order to satisfy these various audiences, the following report format, or a variant of it, is often used that is consistent with the pyramid method of investigation noted previously. This format is based on the classical style of argument used in the Roman Senate almost 2,000 ago to present bills. As it did then, the format successfully conveys information about the matter to a varied audience, who can chose the elevation of detail they wish to obtain from the report by reading the appropriate sections.

- *Report Identifiers*: This includes the title and date of the report, the names and addresses of the author and client, and any identifying information, such as case number, file number, and date of loss. The identifying information can be easily incorporated into the inside address section if the report is written as a business letter. Alternately, the identifying information is sometimes listed on a separate page preceding the main body of the report. This allows the report to be separate from other correspondence. A cover letter is then usually attached.
- *Purpose*: This is a succinct statement of what the investigator seeks to accomplish. It is usually a single statement or a very short paragraph. For example, "To determine the nature and cause of the fire that damaged the Smith home, 1313 Bluebird Lane, on January 22, 1999." From this point on, all parts of the report should directly relate to this "mission statement." If any sentence, paragraph, or section of the report does not advance the report toward satisfying the stated

purpose, that part should be edited out. The conclusions at the end of the report should explicitly answer the question implied by the purpose statement, e.g., "The fire at the Smith house was caused by an electrical short in the kitchen ventilation fan."

- *Background Information*: This part of the report sets the stage for the rest of the report. It contains general information as to what happened so that the reader understands what is being discussed. A thumbnail outline of the basic events and the various parties involved in the matter is included. It may also contain a brief chronological outline of the work done by the investigator. It differs from an abstract or summary in that it contains no analysis, conclusions, or anything persuasive.
- *Findings and Observations*: This is a list of all the factual findings and observations made related to the investigation. No opinions or analysis is included: "just the facts, ma'am." However, the arrangement of the facts is important. A useful technique is to list the more general observations and findings first, and the more detailed items later on. As a rule, going from the "big picture" to the details is easier for the reader to follow than randomly jumping from minute detail to a big picture item and then back to a detail item again. It is sometimes useful to organize the data into related sections, again, listing generalized data first, and then more detailed items later. Movie directors often use the same technique to quickly convey detailed information to the viewer. An overview scene of where the action takes place is first shown, and then the camera begins to move closer to where things are happening.
- *Analysis*: This is the section wherein the investigator gets to explain how the various facts relate to one another. The facts are analyzed and their significance is explained to the reader. Highly technical calculations or extensive data are normally listed in an appendix, but the salient points are summarized and explained here for the reader's consideration.
- *Conclusions*: In just a few sentences, perhaps even just one, the findings are summarized and the conclusion stated. The conclusion should be stated clearly, with no equivocation, using the indicative mode. For example, a conclusion stated as "... the fire could have been caused by the hot water tank" is a guess. It implies that it also could have been caused by something other than the hot water tank. Anyone can make a guess. Professional investigators make conclusions.

As noted previously, the conclusions should answer the question posed in the "Purpose" section of the report. If the report has been written cohesively up to this point, the conclusion should be already obvious to the reader because it should rest securely on the pyramid of facts, observations, and analysis already firmly established.

- *Remarks*: This is a cleanup, administrative section, which sometimes is required to take care of case details, e.g., "the evidence has been moved and is now being stored at the Acme garage" or "it is advisable to put guards on that machine before any more poodles are sucked in." Sometimes during the course of the investigation, insight is developed into related matters that may affect safety and general welfare. In the nuclear industry, the expression used to describe this is "extent of condition." Most states require a licensed engineer or professional to promptly warn the appropriate officials and persons of conditions adverse to safety and general welfare to prevent loss of life, loss of property, or environmental damage. This is usually required even if the discovery is detrimental to his own client.
- *Appendix*: If there are detailed calculations or extensive data relevant to the report, they go here. The results of the calculations or analysis of data are described and summarized in the "Analysis" section of the report. By putting the calculations and data here, the general reading flow of the report is not disrupted for those readers who cannot follow the detailed calculations or are simply not interested in them. And for those who wish to plunge into the details, they are readily available for examination.
- *Attachments*: This is the place to put photographs and photograph descriptions, excerpts of regulations and codes, lab reports, and other related items that are too big or inconvenient to directly insert into the body of the report, but are nonetheless relevant. Often, in the "Findings and Observations" portion of the report, reference is made to "Photograph 1" or "Diagram 2B, which is included in the attachments" due to space limitations.

In many states, a report detailing the findings and conclusions of an engineering investigation is required to be signed and sealed by a licensed professional engineer because by state law engineering investigations are the sole prerogative of licensed, professional engineers.

Thus, on the last page of the main body of the report, usually just after the "Conclusions" section, the report is often signed and dated, and the report is sealed by the responsible licensed professional engineer, or engineers, who performed the investigation. Often, the other technical professionals who worked under the direction of the responsible professional engineer are also listed, if they have not been noted previously in the report, or do not have their own section in the report.

Some consulting companies purport to provide investigative technical services, investigative consulting services, or scientific consulting services. Their reports may be signed by persons with various initials or titles after their names. These designations have varying degrees of legal status or legitimacy vis-à-vis engineering investigations, depending upon the particular state or jurisdiction. Thus, it is important to know the professional status of the person who signs the report. An engineering report signed by a person without the requisite professional or legally required credentials in the particular jurisdiction may lack credibility and perhaps even legal legitimacy.

In cases when the report is long and complex, an executive summary as well as perhaps a table of contents may be added to the front of the report. The executive summary, which is generally a few paragraphs and no more than a page, notes the highlights of the investigation, including the conclusions. A table of contents indicates the organization of the report and allows the reader to rapidly find sections and items he wishes to review, and warns him away from the materials he does not want to review.

Three Sample Reports

Three sample reports are provided following this section. The first is about the investigation of a trash fire that occurred in the machine shop of a nuclear power plant. The consequence of the fire was that a notification of unusual event (NOUE) was declared at the plant. The second is about the investigation of an explosion in a small, coal-fired power plant during start-up. The explosion caused considerable damage to the plant and significantly delayed the plant from going back online. The third is about a specific component failure, a bearing failure, within a multistage, high-pressure centrifugal pump. Since the pump system had sufficient redundancy to continue operations without a problem, the failure of the pump did not cause any significant consequences.

To fit the client's particular needs, the first report was organized as follows:

Cover page: Title and identifiers (Figure 6.1).
Abstract: One-paragraph summary of the report.
Purpose: The question the investigation should answer.
Event Scenario: Short overview that tells the story of this failure.
Root Cause and Causal Factors: Explanation of why the event happened; provides a succinct answer to the question posed in the "Purpose" section.
Risk and Extent of Condition: Where else in the plant this event could occur based upon what has been learned from this event.
Immediate and Interim Compensatory Actions: What is being done presently to control the risks identified in the "Risk and Extent of Condition" section.
Longer-Term Corrective Actions and Enhancements: What is being done to fix the problem.

Appendix A: Fire Scene Investigation: The various findings and observations made during the investigation of the event (Figures 6.2 through 6.7), their documentation, and an assessment of their significance. This contains basis information for the root cause and causal factors.
Appendix B: Operating Experience Information: Summary of what was discovered when the history of the problem was researched. This provides basis information for the root cause and causal factors.
Appendix C: Failure Mechanics and Risk: A discussion of the technical and scientific underpinnings that explain how, why, and when the event can occur, and a discussion of how to prevent it from happening again. This provides basis information for the root cause and causal factors.

The specific order of the first sample report varies from the Roman Senate model outlined in the previous section. However, it contains all the elements of that model and accomplishes the primary goal of that format—of providing information to satisfy the requirements of several types of readers.

The first sample report, which is about the investigation of a trash fire in a mechanical repair shop of a nuclear power plant, has been edited from the original version to save space. However, it still contains all the salient features related to reporting the results of a failure investigation.

The unedited version, which runs about fifty typewritten pages, includes verbatim interviews; cause codes used to administratively track failure types in a computer database; verbatim product warning and information notices; items relating to administrative and regulatory reporting obligations associated with the event; a section on cost estimates associated with fixing the problem; a specific detailed list of corrective actions, which departments in the company will do them, and the schedule when they will be done; and other industry-specific and plant-specific information. The inclusion of all this other material, some of which only indirectly relates to the event being investigated, is typical of nuclear industry reports, which often try to include all related issues within a single report.

The second report, which has been edited to remove identifiers, follows more closely the Roman Senate format described in the first section of this chapter. It is organized as follows:

Cover page: Title, identifiers, and one-paragraph abstract.
Purpose: The question the investigation should answer.
Background Information: Information about the plant and a brief paragraph of the circumstances when the event occurred.
Findings and Observations: A narrative of events going on before and when the explosion occurred (timeline), factual observations about

the damages and conditions existing after the explosion, and information recorded by instrumentation about conditions in the plant when the explosion occurred (Figures 6.8 and 6.10).

Analysis: Evaluation of the verifiable facts and observations related to the event in combination with applicable engineering principles and facts that relate to the event.

Accident Scenario: A succinct chronology or reconstruction of the accident that explains what happened and how it happened.

Conclusion: A two-sentence summary of the fundamental cause.

Epilogue: A short section that provides thumbnail descriptions of some of the organizational and operational shortcomings that set the stage for the event to occur. A note is included to indicate that the details of these shortcomings and how they will be fixed are to be addressed in a separate report.

References: Sources of arcane facts and information used in the report that are in addition to general engineering facts and principles.

Attachment: General overview schematic of the area in the power plant where the explosion occurred.

The differences between the first two case study reports demonstrate well some of the fundamental differences between organizations as to what is expected from an investigation report.

The first report, especially the full, unedited version not shown here, is detailed and inclusive. It not only addresses the technical matters related to the cause of the event itself, but also includes postevent administrative matters, such as what corrective actions, both technical and administrative, are to be done to fix the deficient conditions that were found, which departments will do them, what the completion schedule will be, what it might cost, what regulatory reporting obligations are be associated with this event that have to be addressed, and so on.

In essence, besides determining the cause of the event, the investigation team also provides company decision-making guidance that would usually result from a technical or managerial policy feasibility study. Consequently, the investigation mandate of the first report can be broadly divided into two parts: what, how, and why the event happened; and what and how the company should correct the conditions that led to the failure to ensure that the event will not recur.

Thus, the first case study report serves not only several audiences, but also several purposes. Not surprisingly, this type of report takes more time, requires more resources, and should, but not always does, have senior policy makers actively participating in the investigation team. Implicit in this type of report is the presumption that the investigating team is not only expert in the investigation side of the assignment, but also expert in developing, planning, and implementing the fixes.

The second case study report focuses more on the failure event itself and explains what, how, and why it occurred. The second case study report fundamentally serves only one purpose. Once the problem areas that lead to the event have been clearly identified, the report is complete. Since there are many combinations of ways in which the deficiencies can be fixed that involve company policy, budget obligations, and internal administrative matters, the investigation team leaves those matters to another team or process.

Implicit in this type of report is the presumption that the investigation team is expert about the investigation side of the assignment and has the mandate to carry out this function. However, the team may not have the expertise or the mandate to develop, plan, and implement any specific fixes. This may be an involved process subject to several administrative elevations of approvals.

The third case study report example is a type that is typically used to investigate and document a specific, narrowly focused problem, usually the unexpected failure of a single component, item, or process. Sometimes these reports are called white papers. The scope of a white paper investigation is narrow because, usually, the consequences of the failure are narrow and less significant.

The report usually has an audience of perhaps just a few people, and the audience is often as knowledgeable, or nearly so, as the expert who has been tasked with the investigation and subsequent preparation of the report. The report may only be circulated within a single department, or perhaps among a small number of departments.

Consequently, the language in the report can be more technical and can dispense with background explanations of information and principles commonly understood by both the preparer and the audience. Not surprisingly, this type of report is less formal and has a more abbreviated format.

It is not unusual, however, to find that the evidence, findings, and conclusions in white papers may later on be incorporated into other, more formal reports. For example, the bearing misalignment concluded to have occurred in the third case study might be symptomatic of an overall knowledge and skill deficiency that has led to several such bearing failures, necessitating a programmatic remedy.

The third case study example included in this chapter is organized as follows:

Title page: Includes identifiers and brief statement of purpose.

Brief Background Information: Just enough information so that the reader knows something about the circumstance of the failure.

Observations: A recounting of the pertinent verifiable facts upon which the report's conclusion is based.

Analysis: The engineering analysis that applies engineering methods and principles to the facts to assess what has happened.

Conclusions: One short paragraph that indicates why the bearing failed, and another short paragraph that recommends how to avoid similar failures.

Root Cause Investigation of Fire in Power Plant Shop Area

COOPER NUCLEAR STATION

Brownville, Nebraska

Randall Noon, PE
Root Cause Team Leader
June 17, 2005

Figure 6.1 Fire-damaged area in multipurpose facility.

ABSTRACT

This report describes a trash fire that occurred at the Cooper Nuclear Station in Brownville, Nebraska, on March 14, 2005. A 1,000-watt high-intensity discharge mercury vapor lamp located in the ceiling of a shop area caused the fire. Due to both the intrinsic properties of the lamp and the way in which the lamp had been operated, the discharge tube within the lamp ruptured and hot quartz particles dropped into a trash bin. The hot particles then ignited flammable trash inside the bin by contact.

PURPOSE

To determine the cause of the fire that occurred in the CNS multipurpose facility on March 14, 2005, and to prevent or mitigate the risk of a similar fire recurring.

EVENT SCENARIO

Sensors detected the fire at 2:47 a.m. on March 14, 2005. The fire was then visually confirmed by operators to be in a plastic trash bin in the multipurpose facility, the inside shop area, at Cooper Nuclear Station and was quickly extinguished. The circumstances that led to the fire are as follows:

- Lamping in the multipurpose facility is primarily accomplished with high-intensity discharge (HID) mercury vapor lamps. On Friday, March 11, 2005, an HID mercury vapor lamp located directly above the plastic trash bin was anecdotally reported to have been flickering.
- The trash bin was empty at 2:30 p.m. on Friday, March 11, 2005. After 2:30 p.m. and during the weekend of March 12–13, 2005, cardboard, wood, plastic, and other trash items were placed in the bin.
- Due to an end-of-life phenomenon called de-vitrification, the HID lamp that had been reportedly flickering sustained an arc tube rupture. Hot pieces of quartz, some perhaps as hot as 1,832°F, dropped from the lamp fixture into the bin. The bin was about 45 feet directly below the lamp.
- During the approximate 2 seconds of fall from the fixture to the bin, the hot debris remained sufficiently hot to ignite paper and plastic materials.
- The ensuing fire melted the plastic sides and bottom of the bin. Burning materials fell out of the bin and onto the floor next to a metal modular shipping container that was adjacent to the bin. These burning materials heated and burned the exterior painted metal wall of the metal shipping container.
- Heat conducted through the metal wall of the metal shipping container and damaged the interior paint, and perhaps some plastic materials leaning against the interior wall. Initially this was a smoulder. When the doors of the container were opened to investigate the smoke, fresh air was admitted and the smoulder ignited.

ROOT CAUSE AND CAUSAL FACTORS

The *root cause* of this fire was the inappropriate use of open-fixture, non-O-type HID lamps over an area that regularly contains a fire load.

The arc tube in HID lamps can unexpectedly rupture when they are run to failure without being periodically turned on and off. Examination of the fire scene confirmed that arc tube rupture of the lamp above the trash bin occurred.

Warnings and advisories on the replacement lamp packaging indicate that either this type of lamping should have a cover over the fixture, to prevent hot, rupture debris from falling out, or a type O lamp and ballast should be used. A type O lamp and ballast is designed to contain arc tube rupture debris.

There were two *contributing causes* for the fire. The first was housekeeping. When the arc tube ruptured, the hot arc tube debris could have either fallen harmlessly onto a noncombustible surface, such as a bare concrete floor, or landed on combustible materials. Due to housekeeping practices, it landed on combustible materials.

The second *contributing cause* was an administrative deficiency concerning keeping up with safety standards that change over time. Safety warnings and advisories concerning HID mercury vapor lamps were published by OSHA as early as 1986. The lamp manufacturer and the National Electric Manufacturers' Association have published additional advisories since then. The newly published 2005 National Electric Code specifically addresses the issue of HID lamps. Warnings about arc tube rupture and the consequential fire hazard are directly printed on the packaging of HID replacement lamp bulbs. Despite this information, the hazard was not recognized. This noncognizance is attributed to the fact that when the building was designed and built, lamping did meet the safety standards in effect at the time. The building's lamping system had not been reassessed since then.

RISK AND EXTENT OF CONDITION

The arc tubes in non-O-type mercury vapor HID lamps can unexpectedly rupture when they are run to failure and are not periodically cycled on and off. In open fixtures, an arc tube rupture creates falling hot quartz debris. If it falls on combustibles, a fire can ensue. If it falls on persons, it can injure them. If it falls into equipment or into items open to the air, such as tanks containing liquids, it can cause damage, fire, or contamination.

A survey of Cooper Nuclear Station found that there were perhaps several hundred 1,000-watt and 400-watt HID mercury and metal halide lamps in use. None of the HID lamps were being cycled in accordance with the manufacturer's recommendations to minimize arc tube rupture risk. No one in facilities or maintenance was familiar with those recommendations.

IMMEDIATE AND INTERIM COMPENSATORY ACTIONS

On March 16, 2005, two days after the fire occurred, the root cause team leader and the electrical superintendent met and discussed actions to ameliorate short-term risk. The actions included reduction of fire load in the area through better housekeeping and immediate replacement of 1,000-watt lamps that are dim, burnt out, or flickering. Operators were instructed to periodically turn off and on HID lamps in accordance with the manufacturer's recommended schedule, and a complete re-lamping of HID lamps was planned.

In the course of re-lamping the multipurpose facility, it was discovered that some of the lamps did not have the correct matching ballast. Further, it was discovered that one of the non-HID mercury vapor lamp fixtures had been damaged by excessive heating by the lamp: its plastic cover was charred and had partially melted. When the HID mercury vapor lamps were initially turned off and on, i.e., cycled, two failures occurred. One was in a lamp that had been noted to be dim. It failed to come back on as expected but the arc tube did not rupture. In a second lamp that was not noted to be dim but had anecdotally been noted to occasionally flicker, this lamp sustained an arc tube rupture.

LONGER-TERM CORRECTIVE ACTIONS AND ENHANCEMENTS

The following is a summary of the corrective actions that were taken:

- All high-bay, HID lamping in the plant that could suffer arc tube rupture was surveyed and evaluated for risk.
- A scheduled preventive maintenance task list was developed and executed to replace the interim actions.
- Site housekeeping awareness was improved.
- A study was initiated to determine what should be done with existing HID lamping at the plant to bring it into compliance with the 2005 NEC, as well as OSHA and good IEEE lamping standards.

Appendix A: Fire Scene Investigation

The root cause team leader and the plant fire marshal conducted a fire scene investigation in the multipurpose facility on March 14, 2005, beginning about 8:30 a.m. The following findings and observations were made:

1. The fire originated inside a gray-colored plastic bin located on the west side of a CVAN, a metal truck shipping container. The CVAN had been set next to a bay door in the middle of the multipurpose facility. The CVAN exhibited fire damage on the exterior side of a metal wall proximate to the plastic bin. The CVAN also had secondary fire damage on the corresponding interior side where heat from the fire had conducted through the steel wall and affected the paint.
2. An obvious V pattern was notable on the exterior of the CVAN. The vertex of the V was along the bottom of the CVAN, near a corner of the plastic bin. At the vertex of the V pattern, the paint on the CVAN was noted to have been burned away and bare metal was extant. In areas removed from this area, damage to the painted exterior was less severe. The damage to the paint on exterior of the CVAN followed the usual pattern of discoloration, bubbling, peel-off and exposure of underlying primer, peel-off of primer and exposure of galvanized metal, oxidation of zinc, and exposure of bare base metal corresponding to exposure to increasing temperatures. Thus, bare metal indicated the location of hottest exposure. In typical fashion, the exposed bare metal, steel, had rusted. Consequently, the rusted areas indicate where exposure to the fire heat was greatest.
3. The plastic in the bin had largely melted and fallen to the floor around the bin and resolidified when the floor cooled it. Thus, there was a rough crust of resolidified bin plastic around the bin, especially on the side toward the CVAN.
4. A check of the contents of the bin and the materials that had fallen to the floor from the bin found that they consisted of cardboard, plastic sheeting,

Figure 6.2 Location of fire area and CVAN in the multipurpose facility.

Figure 6.3 V fire pattern on CVAN. Metal frame of bin in foreground.

miscellaneous plastic items, two large blocks of wood, and some dirty cotton shop rags. No containers that may have contained flammable liquids were found. No smoking materials were found. (It is a nonsmoking area that is rigorously enforced.) The cotton shop rags were not oily, and there were no indications of spontaneous combustion in the rags due to reactions with paint thinner, tung oil, or similar. Fire markings on the CVAN and nearby metal bin indicate that the plastic bin was at least three-quarters or more full of fire load materials when it burned.

5. The high-energy, mercury vapor lamp directly above the bin was noted to be blown out; that is, the bulb was broken apart and only a portion of it remained in the ceiling fixture. A pattern of bulb envelope glass was noted around the bin, including glass shards on top of the CVAN, nearby toolboxes, tables, and other surfaces. The shape, type, and thickness of the glass pieces indicated that they had originated from the mercury vapor lamp. It was noted that all the glass shards that were undisturbed by fire fighting efforts were dirty on the topside, but clean on the underside. Further, the surface upon which the glass shards lay was clean underneath but dirty with fire debris elsewhere. The effect is sometimes called shadowing.
6. The mercury vapor lamp was about 45 feet above the plastic bin. Neglecting air friction, the fall time from the lamp to the bin is about 1.67 seconds. This is not enough time for glowing arc tube materials to cool. Consequently, any hot arc tube materials from the lamp that fell into the bin were sufficiently hot to ignite fire load in the plastic bin.
7. It was reported in a statement that the lamp that had blown was observed to be flickering on the Friday before the fire. Flickering indicates that the lamp is at the end of its service life. Darkening or discoloration of the lamp

Figure 6.4 Trash bin exemplar.

indicates it has reached the end of its service life. Product literature indicates that the lamps should be replaced before they fail because they have a higher probability of arc tube rupture at the end of service life. Failure often damages the ballast also.

8. Examination of the lamp found that it is a 1,000-watt Sylvania brand, type H36GV-1000-DX R. The transformer, or ballast, for the lamp is rated at 480 volts. Product literature indicates that the rated average life is 24,000 hours, that is, 2 years, 8 months, and 26 days. The approximate initial lumens are 57,500, and the CCT is 6,300 K. There were no screens or covers on the bottoms of the fixtures in the multipurpose facility as is recommended in the 2005 National Electric Code, Section 410.101(A)(5). Examination of the lamp fixture found that it had not been affected by smoke or water from the fire and had not been hit by the stream of water. Product literature indicates that until the environmental temperature is very hot, over 400°F, the lamps are unaffected.
9. Examination of the bulb that had broken above the plastic bin found that
   a. the black glass portion of the socket (below the threads) was cracked,
   b. the bottom electrical conducting pad or contactor had two noticeable areas of arcing,
   c. the threaded portion of the socket had carbon tracking along the threads, and
   d. the harp that helped support the quartz inner arc tube was temperature affected in the area where the quartz had blown out.
10. The box in which replacement lamps come has a warning on it. The warning states "that the bulb can unexpectedly rupture due to internal causes or external factors such as a ballast failure or misapplication. An arc tube

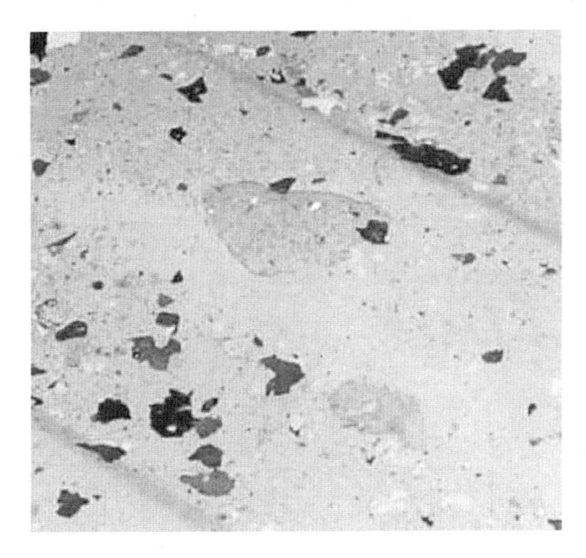

Figure 6.5 Example of glass shard on top of the CVAN with dirt on its topside but clean underneath. This shard shadowed the area under it. The shard matches the type of glass used in the bulb's outer envelope.

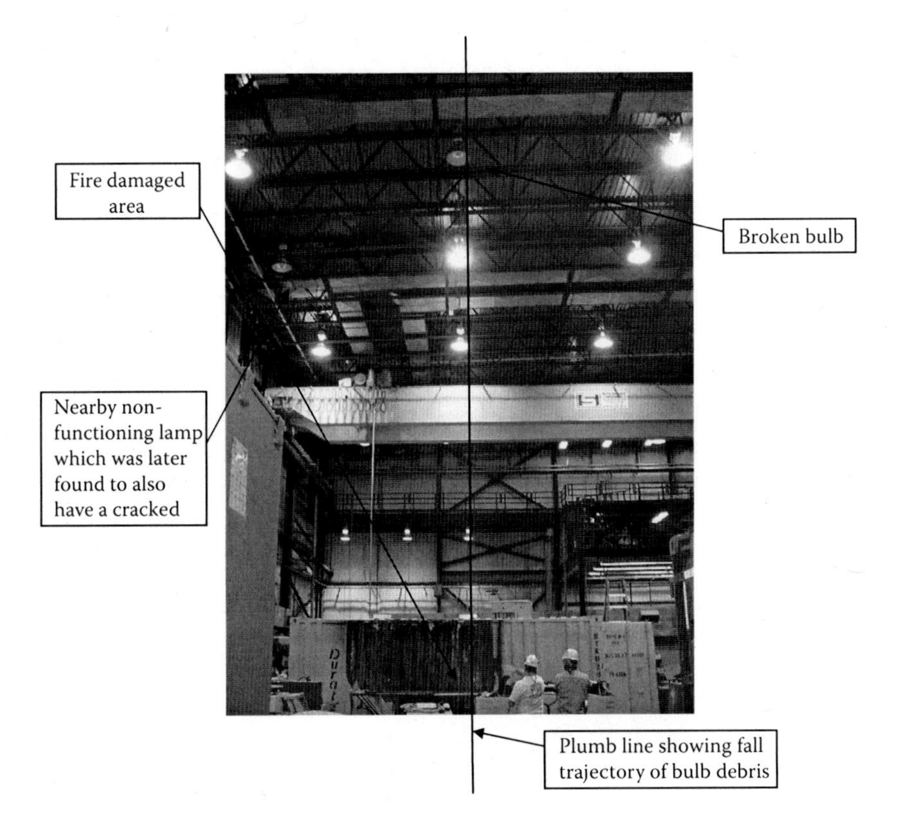

Figure 6.6 Relation of lamp to fire area.

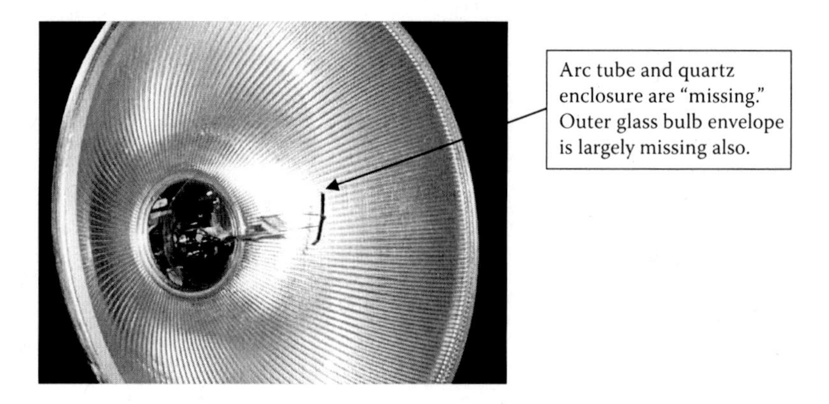

Figure 6.7 Broken bulb as observed in its fixture in the MPF ceiling.

can burst and shatter the outer glass bulb resulting in the discharge of glass fragments and extremely hot quartz particles (as high as 1,832°F). In the event of such rupture, THERE IS A RISK OF PERSONAL INJURY, PROPERTY DAMAGE, BURNS, AND FIRE."

11. A second lamp, which was noted to have burned out but was not ruptured, was also removed from the ceiling area. This lamp was located near the one that ruptured and is visible in the photograph with the plumb line. It is in the row of lamps directly behind the one that ruptured, on the left side of the lamp that ruptured. The outer glass envelope of this second lamp was clouded over, presumably by oxidized arc tube materials. Further, cracks in the black glass base, similar to those observed in the ruptured bulb, were noted.
12. A check found that there is no schedule for bulb replacement. Bulbs are replaced on an as-needed basis when they fail. This is true for all areas in the CNS plant where the HID lamps are used. The manufacturer recommends that the lamps be changed out before they fail to avoid damaging the ballast and to avoid stressing the bulb such that it might rupture.

From the above data, both investigators concluded that the fire was caused by failure of the high-intensity discharge, 1,000-watt lamp bulb located directly above the bin in which the fire occurred. The lamp bulb failed and ruptured, and hot arc tube materials dropped into the bin. The available fire load in the bin then caught fire.

Appendix B: Operating Experience Information

When CNS was built, mercury vapor high-intensity lamping in the high bay areas of the plant was compatible with the National Electric Code and other applicable standards. A check of the 1984 NEC, for example, found no prohibition related to HID lamping except when flammable gases might be in the area.

In 1986, OSHA noted that lamps of the metal halide type could experience unexpected rupture of the arc tube. A public hazard information bulletin was issued then, which is still available on the Internet.

By 1988, Underwriters Laboratory had taken note of the fact that sometimes high-intensity discharge (HID) lamping occasionally sustained internal rupturing of the arc tube. Standard UL1572 was revised to require that lamps meeting this particular specification be able to withstand arc tube rupture.

By 1997, there were reports that the quartz inner arc tube of HID lamps might unexpectedly rupture and cause fires, burns, and property damage. Consequently, warnings had been placed on the packaging of high-intensity discharge bulb replacements. Some agencies published advisory articles, such as the Pacific Energy Factsheet, that noted HID lamps sometimes sustained arc tube ruptures, especially when the lamps approached their end-of-life service time and had been in continuous use.

Beginning in 2000, NEMA published several advisory white papers relating to arc tube rupture. OSRAM-Sylvania followed suit and also published a number of product warnings concerning mercury vapor R lamps.

In 2005, the National Electric Code incorporated specific rules concerning the use of such lamping. Essentially, the 2005 NEC requires that, if applicable, such lamps either be the O type, which can withstand an arc tube rupture, or provide guards and barriers that prevent hot rupture debris from dropping down onto flammable materials. With respect to CNS, there are no O type direct bulb replacements for mercury vapor lamps. Changing to O type also requires a change in the ballast.

Appendix C: Failure Mechanics and Risk

ARC TUBE FAILURE MECHANICS

In high-intensity discharge (HID) lamps, such as the 1,000-watt R mercury vapor lamp in this case, the plasma arc generated within the quartz arc tube produces lamp and heat. High-voltage electricity within the arc tube flows from one electrode, through a pressurized gas medium, to another electrode on the other side.

This is why this type of lamping requires a higher than normal voltage supply, and why it requires a ballast. The ballast performs two functions:

1. It provides a higher than normal voltage needed for the initial strikeover to occur and establish an arc, and
2. It then provides the normal, somewhat lower operating voltage that maintains the arc once it has been established. Essentially, the ballast is a small, two-voltage transformer.

High-intensity discharge (HID) lamps fail for many reasons, of course, including the usual ones associated with poor handling, breakage, misapplication, poor installation, vibration, and so on. The failure mode in this specific case, however, is due to de-vitrification. This is an end-of-life phenomenon.

De-vitrification is when, over time, the plasma arc within the pressurized quartz arc tube slowly ablates the inner quartz walls of the arc tube. In other words, the plasma arc thins the walls of the arc tube by slowly vaporizing some of the quartz. Consequently, the stresses in the quartz arc tube, due to the internal pressure exerted by the ionizing gas that fills it, increase. Thus, when the arc tube has sufficiently thinned, it can unexpectedly rupture. Because quartz is brittle, it does not yield, stretch. or distend significantly prior to stress failure.

The arc tube may also rupture in response to small transient stresses caused by the lamp flickering, strikeover, vibrations, or similar. While de-vitrification may not have proceeded sufficiently to cause rupture by itself, additional transient stresses may combine with the existing elevated stress elevations to exceed the rupture strength of the arc tube.

The normal, end-of-life failure mode for this bulb is a slow outgassing of pressurized mercury vapor from the arc tube. As gas leaks from the pressurized arc tube, the starting voltage required to accomplish strikeover increases. This is evident by an obvious increase in start-up time when the lamp is first energized, or by dimming. Older lamps produce noticeably less light. At some point, the loss of pressurized gas from the arc tube is sufficient so that there is insufficient voltage available from the ballast for the lamp to accomplish strikeover. At this point the lamp may dimly flicker or simply fail to strikeover and light.

This is why unexpected arc tube rupture primarily occurs when the lamp is near its end-of-life service time. As noted in product literature, the rated life of this bulb, whether it is 1,000 watts, 400 watts, or 250 watts, is 24,000 hours. This is equivalent to continuous use for 2 years, 8 months, and 26 days. At the rated life, as much as 50% of the lamp population will have failed according to the manufacturer. Consequently, the rated life should be considered the product's mean time to failure under typical conditions.

When HID lamps such as these are run to failure without being periodically turned off and on, their arc tubes can fail unexpectedly; de-vitrification continues until thinning of the quartz tube is sufficient for rupture to occur. Although a lamp may have leaked enough gas such that it cannot start, it may still have enough gas in the tube to maintain a plasma arc if it is continually energized. Maintaining arc requires less voltage than initiating arc. If a lamp is turned on and off periodically, however, and the lamp eventually fails to achieve strikeover and no longer lights, the process of de-vitrification stops. The risk of unexpected arc tube failure is therefore obviated.

HID lamps that are run to failure also have a relatively high incidence of ballast failure. It is common to find a burned-out bulb together with a burned-out ballast.

RISK

There are three types of risk associated with this type of lamp fixture: fire and explosion, introduction of foreign material into sensitive systems, and personal injury.

With respect to environmental concerns, as long as the mercury vapor lamp is intact there are no known health hazards. Further, no adverse effects are expected from occasional exposure to broken lamps. However, as a matter of good practice, the product safety sheet recommends that prolonged exposure or frequent exposure to broken mercury vapor lamps should be avoided unless there is adequate ventilation. This is to prevent accidental inhalation of vapors.

As noted in the previous section, the risk of unexpected failure of an HID lamp is maximum when older type HID lamps are run to failure and are not periodically turned on and off. This had been the practice at CNS prior to March 14, 2005.

By turning HID lamps on and off occasionally, the lamps that are near their end of service life do not turn back on. If some of these lamps have thinned arc tubes, they will not be further subjected to transient stresses associated with flickering or a hard start. Thus, they fail to start before they fail from arc tube rupture.

Risk can be further minimized if the older type HID lamps are replaced prior to their end-of-life failure through a preventive maintenance program. Since it is known that perhaps 50% of the lamps fail after 24,000 hours of service, replacing lamps when they have sustained perhaps 20,000 hours of service will lessen the risk of unexpected arc tube failure. Replacement of lamps that have noticeably dimmed also reduces risk of unexpected failure. Dimming is a symptom that a lamp is approaching its end-of-life service time.

Risk can be almost eliminated if newer O type lamps replace the older type HID lamps. This type of lamp is designed to contain an arc tube rupture. Unfortunately, there is no one-for-one O type lamp replacement for HID mercury vapor lamps. If a type O lamp is substituted, an O type ballast has to be also substituted. Alternately, risk can be almost eliminated if the HID lamp fixture is equipped with a guard or barrier that prevents debris from an arc tube rupture falling onto combustible materials.

These two methods, replacement by O type lamps and ballasts or the installation of barriers to prevent hot arc tube debris from falling, are both recommended in the 2005 National Electric Code, Article 410.101(A)(5).

Coal Dust Explosion at 230-Megawatt Coal Plant

ABSTRACT

This report explains how a coal dust explosion occurred at a 230-megawatt coal-fired power plant in the continental United States. The explosion occurred during start-up of one unit of the two-unit plant, which was being put back online after a scheduled outage. No one was hurt in the explosion. However, the precipitator, the bag house, and portions of the boiler, especially the area around the economizer, were damaged. Direct costs to repair the unit were estimated to be $6 to 8 million, and one unit of the plant was down for 3 months for repairs. The cause of the explosion was a combination of equipment and procedural deficiencies.

PURPOSE

This is a case study of how a coal dust explosion occurred in a medium-sized coal-fired power plant located in the continental United States.

BACKGROUND INFORMATION

The generating capacity of the plant is 230 net megawatts from two units. The stack height of both units is 178 feet. The full coal capacity of unit 1 is 66 tons per hour, and the full coal capacity of unit 2 is 78 tons per hour. The unit 1 boiler was put online in the early 1960s. Unit 2's boiler was put online in the late 1960s.

The fuel normally used is sub-bituminous, low-sulfur western Wyoming coal that is mined from the Powder River Basin area. For start-up, however, sufco coal is used. Sufco has a slightly higher heating value of 11,400 BTU/pound. The coal piles are not covered and are exposed to the weather.

The explosion occurred during start-up of unit 1. The start-up followed a scheduled spring outage. No one was injured in the explosion. The explosion, however, caused approximately $6 to 8 million of damage, and the unit did not operate for several months while repairs were made.

FINDINGS AND OBSERVATIONS

The explosion occurred on a regular workday just after the spring outage at 9:25 p.m. local time. Prior to the start-up, it had rained and the coal pile was wet.

According to the incident report, the X coal feeder was started at 7:20 p.m. At 7:30 p.m. operators reported "bridging" and plugging in the X feeder. Because of that, operators started the Y feeder at 7:31 p.m. Operators finally established a consistent coal flow at 8:08 p.m. in the X feeder.

A bucket sample of processed sufco coal was obtained from the start-up pile. It had the same moisture content as the coal used on the day of the explosion. The processed coal could be manually compressed into a softball-sized ball, and the ball easily retained its shape even when tossed gently. It was obvious that when wet, the processed coal had a high degree of cohesion.

Operators had problems with ignition throughout the start-up process and never established a good coal fire. The igniter in Y was shifted over to X. Despite several attempts, no continuous coal fire was established until 8:59 p.m., and this was not a good fire. The operator reported that the fire intermittently flashed and "went south."

An operator reported that an "excellent fire" in the boiler was finally established at 9:18 p.m. The temperature in the boiler began to increase then better than at any previous time during the start-up.

The boiler was purged of ignition natural gas twice prior to the explosion. The first purge began at 8:34 p.m. and lasted 6 minutes. The second purge began at 8:44 p.m. and lasted for 7 minutes.

Sometime after 9:18 p.m., an employee reported that he observed black smoke leaking from a side door at the southeast corner of the precipitator. Operators decided that the coal feed would be held steady until a replacement door was found. After that, fuel flow would be secured and the boiler purged so that the door could be replaced.

Between 9:18 p.m. and when the explosion occurred at 9:25 p.m., an operator reported that when securing the X cyclone, the tertiary air damper was discovered stuck in the closed position. An instrument and control technician was dispatched to fix the damper. The technician worked on the damper while the fire was in the Y cyclone and fixed the problem. When he fixed the problem, he cycled the damper open and closed. The technician reported that about 30 seconds after he cycled the damper open and closed, the explosion occurred.

Within 4 minutes after the explosion, the igniter gas was secured and the forced draft fan was secured. A head count was conducted and all personnel were accounted for. Operators then secured unit 1.

During an inspection walkdown after the explosion, the author noted unburned coal dust within the economizer, in the discharge area of the economizer, and in varying amounts all the way to the bag house. Within the bag house he observed a light layer of unburned coal dust on top of previously deposited ash. The ash is typically a tan or light brown color; the unburned coal dust is black.

Shortly after the walkdown, slowly increasing elevations of carbon monoxide were detected in the air pathways downstream of the cyclone furnace burners, particularly in the pathway just preceding the economizer area. Processed coal smoulders were still being found and extinguished within the flue gas pathway.

The most severe structural damage occurred at elevation 1513, which is downstream of the economizer, shown in Figure 6.8. Elevation 1513 is also noted on the attached schematic of the boiler. This is where structural beams and members

Figure 6.8 Damage around the economizer of unit 1.

Figure 6.9 Overpressure damage to bag house.

deflected the most. Below this area, the author observed that surfaces normal to flue gas flow had been blown downward; above this elevation, the author observed that surfaces had been lifted and dropped down. Lateral damage, that is, damage normal to the flue gas flow, was severest in this same area and lessened in severity either upstream or downstream from this area.

The resulting pressure wave of the explosion traveled downstream into the precipitator and then into the bag house. The pressure wave rebounded in the bag house and caused the bag house to sustain some structural damage (Figure 6.9). Along the way, the pressure wave also distended some expansion joints.

After the explosion, no one reported residual smell of natural gas odorant in the area where the explosion initiated.

The operating parameters of the plant leading up to, during, and after the explosion were reviewed, especially Y and X coal flow, SO furnace temperature, NO furnace temperature, opacity, total airflow, Z cyclone total air, Y cyclone total air, and X cyclone total air. The following items were noteworthy:

- At the time of the explosion and slightly preceding it, the opacity of the flue gas near the stack exit was 100%. From 9:20 to 9:25 p.m., the opacity was near or greater than 80%. During the same time that the opacity was increasing from 80% to 100%, the south furnace temperature and the north furnace temperature were rising.
- During the two purges, the first at 8:34 p.m. and the second at 8:44 p.m., the opacity significantly dropped. Between purges, the opacity jumped up from 25% to about 55%.
- From 9:00 p.m. to the time of the explosion, 9:25 p.m., the total airflow was about 240,000 pounds per hour.

ANALYSIS

The explosion was not caused by the ignition of combustible gas. The two successive purges that totaled 13 minutes were more than sufficient to disperse any accumulated unburned igniter gas within the flue gas pathway. A simple airflow calculation indicates that 13 minutes is more than enough time for all the air in the boiler flue gas pathway to be replaced several times with fresh air.

Further, no residual mercaptan odorant smell was noted in the areas heavily damaged by the explosion. Any igniter gas that might have collected in these enclosed areas would have had enough residence time for odor to be retained in the fibrous insulation materials.[2]

Operators and personnel did not observe any flashback of a fire front or fireball typical of a gas deflagration. In gas deflagrations, the fire front typically follows the fuel trail from the point of ignition back to the source. The author observed no singeing, burn damages, burn "shadows," or similar fire front effects between the point where the explosion was most severe, i.e., the greatest overpressure, and the cyclone furnace area. There was no continuous, explosive elevation, gas trail from elevation 1513 to the cyclone furnace.[2]

As previously noted, the point of ignition was near or at elevation 1513. Damage was most severe at elevation 1513 as evident by bowing and displacement of structural beams and members. Typically, the point of ignition is near or at the location where the deflagration pressure is greatest. When the structure is more or less regular, this means that the point of ignition is near, or at, the location where structural damage is greatest.

The ability of the boiler to confine the deflagration overpressure was limited. The flue gas pathway was essentially a long rectangular tube with both ends unsealed and open to the atmosphere. High, localized overpressure easily blew out panels between structural beams and members. Consequently, the overpressure front resulting from the explosion rapidly diminished in effect with distance from the point of ignition. Because the boiler flue gas path could not effectively contain the overpressure, there was no displacement of the point of heaviest damage from the point of initial ignition; they were coincident.[2]

The fuel source for the explosion was processed coal dust. Unburned coal dust was found throughout the flue gas pathway from the cyclone furnace to the end of the bag house. Just prior to the explosion, instruments indicated that the opacity at the stack was 100%. This indicates that air exiting the stack was heavily laden with black coal dust. Further, there was an eyewitness report that black smoke was observed exiting through the door on the southeast corner of the precipitator. This door is located directly downstream of where the explosion center was located. The eyewitness observation corroborates that just prior to the explosion, the air in the vicinity of the economizer was laden with unburned, suspended coal dust.

The following related items are excerpted from reference 1, pp. 218–219 and 224:

For high volatile bituminous coals, the minimum explosive concentration lies between 50 and 100 g/cubic meter. The upper explosive limit is not well defined, but is above 4000 g/cubic meter. (This last figure equates to about 8.8 pounds per cubic meter.)

- Prevention of coal dust explosions by the reduction of oxygen does not occur until the oxygen elevation is less than 13%.
- The high temperature of drying air often can provide enough heat to ignite coal accumulation....
- The minimum ignition temperature of a coal dust cloud can range from 425 C for dried lignites to 800 C for certain anthracite coals. Bituminous coals typically ignite somewhere between 500 C and 625 C. The minimum ignition temperature for the generation of a smouldering fire in a coal dust layer

is much lower, however, and certain bituminous coals may be ignited at temperatures as low as 160 C.

- The smaller the particle size, the greater the exposed surface area and the tendency to undergo spontaneous heating.
- Volatile gases resulting from thermal decomposition when the coal is heated can cause a serious explosion in a confined area such as a silo or surge bin.

At standard temperature and pressure (STP), 240,000 pounds of air per hour, the amount of air being supplied at the time of the explosion, is equivalent to a volumetric flow rate of 3,137,000 cubic feet per hour, or 52,287 cubic feet per minute.

The cross-sectional area around the cyclone is 336 square feet, and the cross-sectional area at elevation 1502 and higher is 1,056 square feet. The approximate airflow velocities in these areas were 2.6 and 1.5 feet/second, respectively, at STP. At temperatures of about 200°F, the airflow velocities would have been about 3.2 and 1.87 feet/second, respectively. At these velocities, it would have taken air approximately 25 seconds or so to travel in a straight lie from the cyclone to the 1513 elevation.

The above corroborates well the time estimate provided by the instrument and control technician: that he heard the explosion about 30 seconds after the tertiary air damper was cycled. Operating the tertiary air damper modulates airflow into the cyclone and can dislodge partially burned coal fuel particles still in residence in the cyclone. Any particles so loosened would have taken that amount of time to travel to the coal dust cloud and ignite it.

ACCIDENT SCENARIO

The evidence indicates that the following occurred.

The coal was wet and sticky. This not only caused handling problems, but also tended to quench combustion in the cyclone. When wet, the coal particles did not mix well with the air in the cyclone; the coal fuel formed large clumps and was heavier than the particles for which the cyclone was designed to handle.

Coal fuel was put into the cyclone, but only a portion of it was burned away by the igniter gas. The rest stayed resident in the system. The two purges that were executed purged the excess natural gas from the system well. This, however, did not purge the excess wet coal fuel that was resident in the system.

The combination of residual heat from start-up attempts and the airflow from start-up attempts and purges finally dried the excess wet coal that had been resident within the system. This coal fuel, once it finally dried out, was then free to be picked up by the airflow and formed a dust cloud that traveled through the system. This is documented in opacity measurements, witness observations, and the fact that large amounts of unburned coal dust were found throughout the boiler system after the explosion. An opacity of 100% is sufficient to support an explosion.

When the instrument and control technician cycled the tertiary air damper, this caused the airflow in the cyclone to modulate. This likely caused a partially burned ember or spark from the cyclone to dislodge and move into the airstream. It then followed the normal flue gas pathway. When it reached elevation 1513, it touched off the explosion.

CONCLUSION

The explosion that occurred was caused by the ignition of unburned coal dust carryover. Ignition difficulties of wet fuel during start-up allowed the excessive carryover to occur.

EPILOGUE

A root cause team, which had been assembled to investigate all aspects of the accident, found the following four key equipment and procedural issues had contributed to the accident:

1. Equipment modifications performed during the spring outage may have affected the firing characteristics of the boiler igniters.
2. An air damper was found to be stuck in the closed position during start-up.
3. Less than adequate operating instructions were available to guide the operators during start-up with the low temperature gains experienced during fuel admission.
4. The failure of the secondary air shut-off damper created a distraction for the operator and resulted in the omission of completing certain operating guide elements.

Subsequent to the above, a burner management team was formed to concentrate solely on addressing the deficiencies found by the root cause team.

REFERENCES

1. *Industrial Dust Explosions,* "Coal Dust Explosions in the Cement Industry," by Amin Alameddin and Steven Luzik, *ASTM STP 958,* Kenneth L. Cashdollar and Martin Hertzberg, eds., ASTM, Philadelphia, 1987, pp. 217–33.
2. *Engineering Analysis of Fires and Explosions,* "Determining the Point of Origin of an Explosion" and "Odorants and Leak Detection," by R. Noon, P. E., CRC Press, Boca Raton, FL, 1995, p. 187.
3. http://www.risktech.se/rdust.htm, short, dramatic video demonstrating a coal dust deflagration. Other dust explosions are also demonstrated.

Evaluation of Bearing Failure in Control Rod Drive Pump A

Randall Noon, PE
Cooper Nuclear Station
August 2, 2001

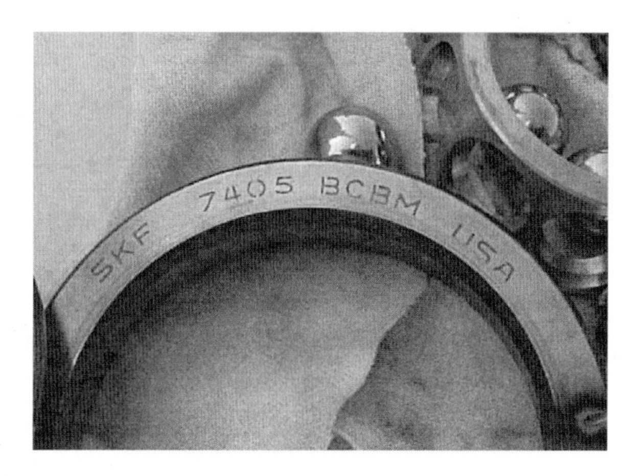

Figure 6.10 Failed CRD outboard bearing.

BRIEF BACKGROUND INFORMATION

The A control rod drive pump failed at Cooper Nuclear Station on August 24, 2001. It failed due to high vibration elevations in the outboard bearing, especially in the axial direction, and poor pump performance.

Output pressure of the pump dropped over a short period of time, and the pump sounded different to the operators who routinely checked it on rounds. Subsequent vibration measurements confirmed that the different sound was due to high vibration elevations generated at the outboard bearing.

The pump failed almost 1 year to the day after the outboard bearing had been field installed.

OBSERVATIONS

The two control rod drive pumps at Cooper Nuclear Station are both rated for 1,600 psig discharge at 79 gpm and 4,210 rpm. They were manufactured by the Worthington Pump Company and are type 2WTF810.

The outboard bearing configuration on the control rod drive pump consists of two ball-type roller bearings installed side by side on the pump shaft. One of the outboard bearings was provided for examination. The rest of this report concerns that particular bearing.

The bearings have a 20° contact angle that is intended to handle both radial and axial loads. From the design of the bearings and the nature of the pump, a multistage high-pressure centrifugal type, the anticipated radial loads carried by the outboard bearing are greater than the anticipated axial loads.

As shown in the cover page photograph (see Figure 6.11), the outboard bearing is type SKF 7405 BCBM. Both races have ten balls per race.

The ball diameter was measured to be about 0.593 inches (15.06 mm). The cage appears to be made of brass or bronze, while the inner and outer races are standard hardened and ground steel. The measured bore diameter is 0.982 inches (24.87 mm). The inner race of the outboard bearing, shown in photograph 1 on the next page (see Figure 6.12), is straw colored along the ball groove and around the bore. This indicates that the bearing had been running at a significantly elevated temperature for some time.

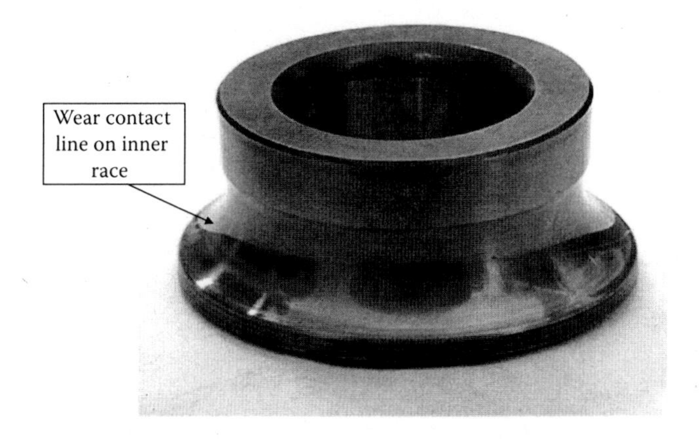

Figure 6.11 Inner race.

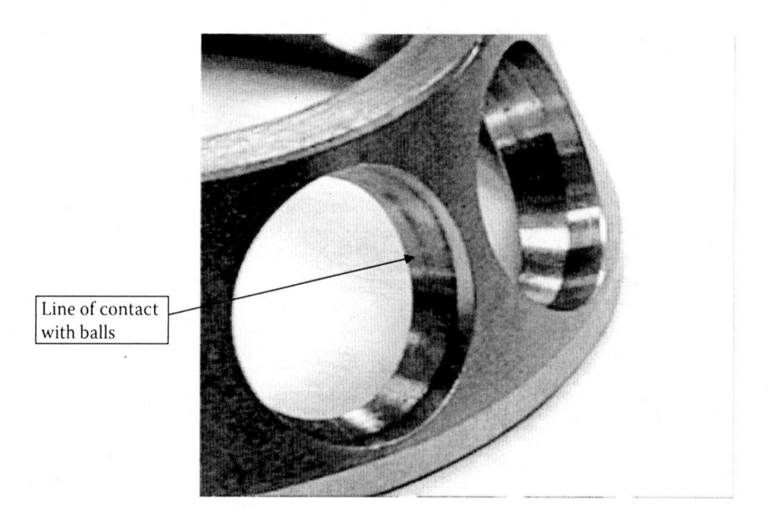

Figure 6.12 Cage of the bearing.

Further, the wear pattern on the inner race where the balls contact the groove is not symmetric; it is tilted. The wear is above the expected contact circle along the ball groove on one side of the race, and 180° away from that location it is below the expected contact circle. This is an indicator of misalignment within the bearing when it was operating.

Examination of the cage under low magnification, shown in Figure 6.12, found that the slots for the balls had heavier than normal wear marks where the balls contact the inner surface of the slots. Further, the wear marks were not symmetric, which also indicates a misalignment present when the bearing was in use.

The outer race, shown in Figure 6.13, was not significantly straw colored. This indicates that it had not yet heated up significantly and that the primary point of heat generation was between the balls and the inner race. Examination of the outside diameter of the bearing found points where fretting between the outer

Figure 6.13 Outer race.

diameter of the bearing and the pump bore was very pronounced, and areas where contact between the outer race and the casing bore was light.

ANALYSIS

The basic equation that describes the life of a roller bearing is given by the following:

(i) $$L = (C/P)^3$$

where
L = rated life in millions of revolutions
P = equivalent load on the bearing, both radial and axial
C = constant associated with the bearing design

With respect to the equivalent load on the bearing, P, there is a second equation that defines that factor:

(ii) $$P = XVF_R + Y F_A$$

where
X = 0.70, factor for this type of bearing
V = 1.0, factor for this type of bearing
Y = 1.63, factor for this type of bearing
F_R = radial load
F_A = axial load

Substituting the equivalent load equation (ii) into the basic bearing equation (i) gives the following:

(iii) $$L = (C)^3/(0.7 F_R + 1.63 F_A)^3$$

Inspection of equation (iii) indicates that when axial forces are present, they decrease the life of the bearing exponentially and have a greater influence on shortening bearing life than radial loads. In fact, a unit of axial load has more than twice the effect on shortening bearing life as a unit of radial load due to the ratio of the coefficients.

For example, consider the case where the radial force is unity, the axial force is zero, and the bearing life is L′. If the axial force becomes just 10% as strong as the radial force, then the life of the bearing becomes 0.53L′. In other words, the expected life of the bearing will be halved.

In this case, a reasonable bearing life for CRD pump bearings is 8 years. This is based upon operating experience at CNS as documented in the CWIT database. If the bearing only lasts 1 year instead, as it did in this case, the equivalent load, P, likely has doubled.

From design considerations, especially the 20° contact angle, it is presumed that under normal circumstances the ratio of radial load to axial load in that bearing is about 3 to 1. Considering the X and Y coefficients in the previous equation, if the radial load remained the same but the axial load increased 3.2 times its typical amount, this would double the equivalent load, P, and shorten the life of the bearing by a factor of 8. If this were to have occurred, it would have produced at the same time a noticeably higher elevation of vibrations in the axial direction.

In this respect, it was reported that when the outboard bearing was installed a year ago, it did initially exhibit higher than expected vibrations in the axial direction. Since the vibrations, however, were considered to be within the acceptable range, the pump was put into service and operated. As previously noted, just before the pump was taken offline, the axial vibrations were noted to be significantly higher than normal. Both of these reports indicate that higher than normal axial forces were occurring in the bearing when it was operating.

Examination of the bearing found that it had been overheating long enough to generally give the inner race an overall straw color. This indicates excessive friction between the balls and the races. Also, wear markings observed in both the bearing inner race and the cage indicate that the bearing inner and outer races were misaligned with each other. Lastly, examination of the outer race's outside diameter (OD) found evidence of significant fretting at certain locations around the OD.

Thus, all of the observations also indicate that higher than normal axial forces were present when the bearing was operating, and that these excessive axial forces were produced by a misalignment in the outboard bearing.

An inquiry was made to determine whether a check for perpendicularly was done after this bearing was installed. Such a check would typically be accomplished using a machinist's dial indicator and magnetic base mounted such that by either spinning the shaft or spinning the bearing, run-out on the bearing can be measured. The inquiry found that the work procedure for field installation of the bearing apparently did not include such a perpendicularly check.

CONCLUSIONS

The outboard bearing of the A CRD pump failed after just 1 year of service because the outboard bearing was out of alignment tolerance. This likely occurred when it was field installed.

This deficiency could have been detected and the premature failure avoided if after installation a simple check for perpendicularity tolerance had been done. It is recommended that this step, or one similar to it, be added to the shop installation procedure for this type of bearing.

REFERENCES

Baumeister, L., and Marks, L. 1967. *Standard handbook for mechanical engineers,* 7th ed. "Bearings with Rolling Contact," NY: McGraw-Hill, pp. 8-178–8-187.

Norton, R.L. 1998. *Machine design: An integrated approach.* Upper Saddle River, NJ: Prentice-Hall, pp. 675–689.

Misplaced Method in the Science of Murder

7

JON J. NORDBY, PHD, D-ABMDI

> Mason said, "There's one nice thing about circumstantial evidence; it's a two-edged sword. It cuts both ways, and it always tells the truth. The trouble with circumstantial evidence is that sometimes we make an incorrect interpretation because we don't have all the evidence."
>
> **Erle Stanley Gardner, *Perry Mason Solves the Case of the Queenly Contestant*, Pocket Books, 1968, p.133**

Introduction

By the way, Erle Stanley Gardner, an amazingly prolific fiction and nonfiction writer and the author of the Perry Mason mystery book series, actually did practice criminal law. Later in life, Mr. Gardner championed the cases of potentially innocent people who had been wrongly sent to prison in a project called the Court of Last Resort.

I am indebted to Jon J. Nordby, PhD, D-ABMDI, for the following excellent case study that constitutes this final chapter. Dr. Nordby is a nationally noted forensic scientist and consultant, the author of *Dead Reckoning: The Art of Forensic Detection,* and the co-editor of the textbook *Forensic Science: An Introduction to Scientific and Investigative Techniques,* which is now in its third edition.

While the three case studies provided previously in Chapter 6 have an obvious engineering slant, in this chapter Dr. Nordby ably narrates a compelling case study about how the same fundamental scientific method and ancillary principles were used to evaluate evidence related to a heinous crime.

The Scene

The state's renowned police expert had seen it all before: experience gleaned from years spent attending to the aftermath of murder. An expert in crime scenes, bloodstains, firearms, and all things to do with homicide, he made his determinations at the scene quickly and concisely after walking through the small single-story house, looking at the ransacked living quarters, seeing the dead woman lying face up on the kitchen floor, and examining the bloodstains he found on the husband's face.

His conclusions were firm and unyielding. The crime scene was staged merely to look like the work of an intruder; the blood on the husband's glasses told him the story of domestic murder. The husband was guilty. He shot his unsuspecting wife in the back with a shotgun sabot slug, sending her into the next world with a single blast from his 12-gauge while she worked at the kitchen sink. The expert gleaned facts sufficient to craft the end of another violent story. He told the photographer to document the scene, as he packed up his notebook and left through the front door.

When he left for the city to write his report, he had been at the rural mountain house about 2 hours, give or take an hour. Only the photographer and three patrol officers, including the first officer to respond to the original 911 call, remained awaiting the deputy coroner. The deputy coroner's chief duty involved collecting and transporting the body for autopsy. The lead detective had taken the prisoner into custody and transported him to jail pending his formal arraignment.

The photographer looked around for something other than the body and the blood to document, like a shotgun, for example. He didn't see one. Maybe someone removed it along with the prisoner. He got the patrol officer's notes and began a hasty review to catch up on the case while he loaded his camera.

The husband had made the 911 call at 7:12 p.m. He called from a neighbor's house, which was about a 5-minute walk or a 3-minute dash from his own. The kitchen phone dangled from one remaining sheet-rock screw still stuck into the wall. The husband told the 911 operator that his wife had been shot—he said he knelt in front of her, felt for a heartbeat, saw she was dead, tried to use the dysfunctional phone, ran to use the neighbor's phone, and called 911 for help.

The operator kept him on the line until the first patrol officer arrived at 7:35 p.m. The officer's narrative had not yet been copied onto the appropriate report form. The photographer looked up from the officer's notebook, toward the kitchen, into the adjoining dining area, and through to the living room/family room. He moved slowly through the small house taking photos as he went.

The dining room table was pushed against the wall with two of its four chairs tipped precariously on spindly legs. The end tables just inside the living room, drawers open, spilled miscellaneous contents onto the beige carpet. The buffet doors were open as were the desk drawers. Papers littered the proximate surrounding flat surfaces.

The home entertainment center displayed empty slots and protruding tangles of wire once feeding now-absent components. An empty holster and a box of 357 cartridges lay beside the open coat-closet door. His flash helped illuminate the details for his camera. He changed rolls of film and moved on through the residence.

The bedroom revealed a neatly made bed that had become a dumping ground for drawers pulled from an oak highboy. Their various contents, including male and female clothing, appeared randomly strewn about the room. The open closet door barely concealed a pile of possessions once suspended from plastic hangers.

The husband had listed various missing property, including audio components, a revolver, a hunting rifle, some jewelry, and some cash from the bedroom. He had said he did not own a shotgun.

The arrival of the deputy coroner provided the photographer with his opportunity to document matters in the kitchen. The body and the kitchen area had already occupied most of the official attention allotted to the case: that, and of course the husband—including his hands, his clothing, and his glasses. The sabot had entered the decedent's back and exited along the midline of her chest.

Three holes appeared through the back of her sweater when the body was moved by the deputy coroner—the sabot had separated as designed, leaving one hole made by the slug, and two holes made by the separating sabot.* The front of her chest was covered with blood.

Bloodstains of varying sizes appeared on the arms and the lower waistband of the once-white fuzzy sweater. The slug and the two halves of the sabot appeared on the counter beside the sink. They would need to be collected and taken into evidence. And a thorough search would have to be made for the murder weapon.

Further Search of the Premises

Searching the property took the better part of three full days. The family truck was found parked near the garage with the passenger door open, and the contents of the glove box strewn onto the bench seat and floor. No shotgun was found anywhere on the property, and none of the items reported to be missing from the house by the husband were found either.

Officers also confirmed that the car pool had dropped off the defendant in front of his house at about 6:55 p.m. the night of the murder.

* A sabot (pronounced *sab bow*) is designed to fire a single large slug through a shotgun barrel rather than multiple numbers of shot contained in a shot-cup. The plastic surrounding the slug is designed to break away in flight after exiting the muzzle. This becomes a secondary missile, which may also strike the target, depending upon the muzzle-to-target distance.

The State's Expert Reconstructs the Murder from the Clues

Like many official reports, the one produced by the state's expert was long on inferences and short on data. The inferences forming the heart of a crime scene or shooting reconstruction remain only as good as the supporting data together with its sound scientific analysis. Both the data and the analysis offered in this report were found wanting.

The report stated unequivocally that the disarray in the small house resulted from the actions of the husband. These actions, the report intoned, were designed to be interpreted as the actions of some unknown intruder who had killed the decedent, robbed the residence, and departed undiscovered.

Instead, the report indicated, they merely exposed the husband's obvious but clumsy plan to cloak his own murderous actions. So the husband shot his wife as she worked in the kitchen, then he proceeded to conceal his nefarious act by pulling out drawers, opening doors, and strewing the household contents around each room. Support produced for this conjecture centered on some inadequately documented broken glass.

The ballistics of the sabot and the bloodstain analysis dovetailed, according to the report, to prove the husband's guilt. The report included a diagram of what purported to be the trajectory of the fatal shot. The expert's assemblage of physical evidence, including the bloodstains on the defendant's glasses, proved that the defendant shot the victim in the back.

When taken into custody, the husband had small bloodstains on his glasses as well as bloodstains on the palms of both hands. Wipe stains appeared over the pockets of his Levis. No other bloodstains of any significance were noted in the report.

The stains on the defendant's glasses were referred to as high-velocity backspatter. The report described the stains on the glasses as being "1 mm or less" in diameter. It further explained that such small stains "can *only* be produced through gunshot impact."

These high-velocity stains result when a bullet strikes a blood source and the resulting impact has sufficient energy to produce small droplets, which fly backward toward the muzzle of the weapon firing the shot. The only mechanism with sufficient energy to produce such high-velocity spatter, at least according to some readers and some writers of the bloodstain pattern analysis literature, is a gunshot.

The fact that these stains appeared on the defendant's glasses offered the only support for the expert's ballistics analysis: he concluded that the shot was fired at a close range—close enough for these tiny droplets to leave the target and strike the defendant's glasses at 90°.

The claim often advanced is that this high-velocity backspatter travels about 20 to 36 inches from the target (blood source) toward the muzzle. The round nature of the stains, the report claimed, supports this angle of impact at 90—when blood drops strike a surface, their angle can be determined by the shape of the stain.

If the drop is round, the angle is 90°. If the drop is oblong, then the length and width of the stain can be used to compute the exact angle of impact according to a well-established formula: angle of impact = $\sin^{-1}(W/L)$, where W = the width of the stain in millimeters and L = the length of the stain in millimeters. The round stains, each 1 mm or less in diameter, absolutely prove that the defendant fired the fatal shot, the report concluded.

In addition, the report emphasized the size of these stains appearing on the glasses—each stain on the glasses was reported to be less than 1 mm in diameter. The scientific claim underlying these conclusions remains unstated: the report's author assumed that only high-velocity bloodstains* created by gunshot could produce such small bloodstains.

The above assumption has been repeated frequently in courts and has gone mostly without question. This assumption joins the others in the report, which remained untested by the expert. Nevertheless, the report concludes by offering a "crime scene reconstruction" based upon the data selected for this analysis.

The reconstruction provided in the report offered an explanation for each of the significant facts of the case: The disarray in the house resulted from the husband's efforts to "stage the scene" to make the death appear to be the result of a robbery gone bad; the bloodstains on the husband's glasses resulted from high-velocity backspatter, which could only come from his firing a shotgun sabot, which struck his wife's back; the husband's presence at the scene simply underscores his involvement in the crime.

When initially questioned by officers and detectives about the inability to find the murder weapon at the scene, the state's expert's reply was, "Absence of evidence is not evidence of absence," a quip some attribute to the patriarch of bloodstain pattern analysis, Herb McDonell. It states that our simple inability to discover or develop specific evidence in a case does not somehow *mean by itself or prove independently* that the evidence was never there to be discovered in the first place. However, inadequate science cannot be covered by clever quips, no matter how they were originally intended. Indeed, the

* According to one bloodstain classification taxonomy, stains are distinguished as being either low velocity (mostly large stains), medium velocity (mostly stains greater than 1 mm in diameter), or high velocity (stains 1 mm or less in diameter) in nature. This terminology, while still used, has lost favor with most scientifically trained bloodstain pattern analysts since it is inaccurate, imprecise, and not clearly descriptive. The energy applied to the blood source to produce blood spatter rather than some vague notion of a droplet's velocity (a vector) is not adequately accounted for by distinguishing low-, medium-, and high-velocity stains.

problem with the scientific work in this case remained the scientific work itself, beginning with the data.

Revisiting the Scene Science: The Problem of Data

Viewing and approaching the scientific project merely in an adversarial light—like two teams clashing on a sporting field with an eye only toward winning the battle of "our results vs. their results"—destroys any potential for the impartial examination of evidence.

Of course, forensic science remains scientific work to be presented in court. This automatically puts the *presentation of scientific results and their assessments* into an adversarial context—a context in which interested parties become very partial. But the practice of science itself, the science applied to the data in order to reach those results, must never become guided by an eye fixed solely toward some desired or favored outcome based upon extrascientific interests. In science, the data should prevail, and the results should fall within the proper limits of the science applied to these data.

One potential adversarial problem involves the close associations among interested police investigators, enthusiastic detectives, and ambitious prosecutors with police scientists working in police crime laboratories. This police influence can, at the very least, appear to be a conflict of interest with respect to the practice of independent data assessments and disinterested applications of natural science.* If it does not become an actual conflict, this fact is due in no small measure to the individual integrity and scientific professionalism of the specific crime laboratory analyst involved in the case. This chauvinistic attitude, unless circumvented by this individual's moral fortitude, can result not only in the refusal to share data but also in the refusal to explain the specific scientific protocols, techniques, and reasoning applied to derive a particular result.

These omissions become evident in official reports that present only conclusions and rarely the work involved in their derivation. In this way, one important element of the scientific method, independent assessment either generating mutual support or raising additional doubts, remains still-

* Some crime laboratory analysts even openly state, "I won't talk with the defense," or they accuse those who do of "going over to the dark side." This hardly reflects a disinterested scientific approach to data that lets the science applied to that data stand openly, allowing other scientists the opportunity to check, test, and either concur or supply a chain of reasoning supporting any differences or disagreements. This notion of potential independent assessment remains an essential part of the scientific method, robbed of significance by any such proprietary, parochial attitude.

born. Scientists hired by the defense can be hampered by the ill will of the prosecutor who retains ultimate control of the evidence.*

This adversarial problem is not unique to those experts working with the prosecution. Of course, a similar problem can arise in the close association between defense attorneys and the experts that they hire. Unfortunately, some experts are more anxious to sell their testimony by making it favorable to defense interests rather than to supply independent scientific analyses of independently developed data.

These ethical issues and the resulting problems for scientific method are by no means novel or exclusive to either the prosecution or the defense. Suffice it to say that ethical lapses, wherever they occur, denigrate the application of scientific method as well as the reputations of those who take ethical shortcuts either for personal gain or for the sake of some independently perceived morally right outcome trotted out to justify such scientific sin.

The intellectual baggage revealed by the state expert in this case provides additional insights into how scientific method, or at least what may superficially pass for it, can go badly wrong. The data remain the focus of the problem.

Asking questions is always preferred to answering questions, at least when it comes to the initial generation of data. When applying science to the investigation of a crime at a crime scene, an unfortunate cookie-cutter approach to the sciences in the expert's tool box can take hold.

In this approach, the data are recognized and collected according to nothing less than specific pseudoscientific prejudices seemingly based upon the statistical frequency of some observation rather than upon the observation of actual data discovered and documented at the scene. Such pseudoscientific generalizations can be confused with covering scientific laws, laws supposedly allowing the deduction of particular cases as logical consequences of these scientific laws.

The data, therefore, become recognized as an instance of the familiar law, and the law then provides the scientific explanation of the data. The potentially inappropriate circular nature of this reasoning remains obvious. One example involves claims about high-velocity bloodstains resulting only from gunshot.

The 1 mm bloodstains on the defendant's glasses were both *recognized* as data through the application of this inferential process (less than 1 mm is high velocity; high velocity is only from gunshot; therefore, this blood on the

* In one case, when the prosecution denied me access to the physical evidence I was to examine, I simply asked the attorney who hired me to ask the prosecution why they bothered to collect, document, and save the evidence if it was to remain locked away, never to see the light of day. Sometimes this jogging of the legal memory is sufficient to stimulate the release of materials for scientific analysis.

glasses is high velocity from gunshot) and *explained* as the result of gunshot through this deductive process.

Scientific methods, however, only begin with the recognition of some data as puzzling: data that requires an explanation. It cannot end with the imposition of an explanation simply imposed at the point of recognition. Of the scientific issues apparent in this case report and expert reconstruction, the high-velocity bloodstains reported on the suspect's glasses remain one of the most informative about scientific method.

Testing to Develop Scientific Inferences from Data

The blood evidence available in this case included some photos of the scene, the decedent's clothing, the suspect's clothing, and the suspect's glasses. Bloodstains can tell us a great deal about any blood-shedding event. Bloodstain pattern analysis, as a discipline, involves the scientific study of the static consequences resulting from dynamic blood-shedding events.

As observations, bloodstains must be described and documented. If equivocal mechanisms could produce any given stain pattern, then testing becomes essential. The tests are designed to help determine, if possible, the mechanism that produced the given stains at the scene. Such questions posed about the stains on the suspect's glasses, and on the decedent's sweater, revealed essential information permitting scientifically supported inferences about the shooting event. These inferences replace the apparently circular inferences resulting from the cookie-cutter approach to applied science described above.*

Scientific Inferences from the Decedent's Sweater

The first task involves observation, followed closely by description. I observed the decedent's sweater, front and back. I noted that the stains on the front,

* This method, if one could call the approach a method, precludes essential elements of what we refer to today as the scientific method. In what I have called the cookie-cutter approach to crime scene data, the science appears in the form of "wise sayings," which function scientifically to inform us about the data. (This can even be taught as a form of deductive reasoning.) Sometimes this approach results in the correct answer, but if one recalls one's early scientific training, it makes none of the correct scientific methods evident. That is why, presumably, we were asked to "show our work" when solving chemistry, physics, or mathematics problems. The right answer, without supporting reasons, is *scientifically meaningless*. As I recall my own scientific education, some students even got the right answer by copying from other students who coincidentally had derived the correct answer. The method of copying, however, is not to be confused with any scientific method of computation being measured through the work.

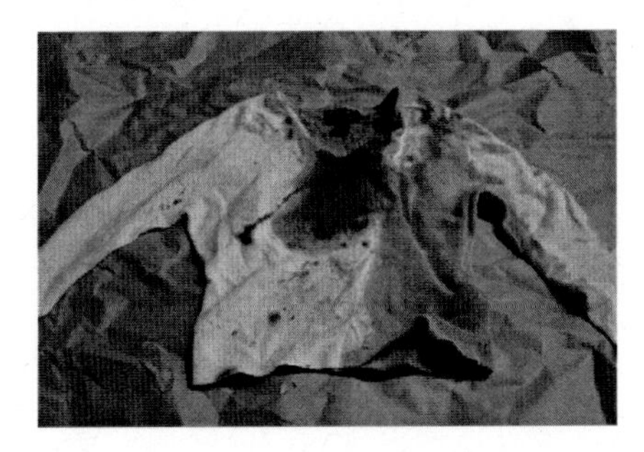

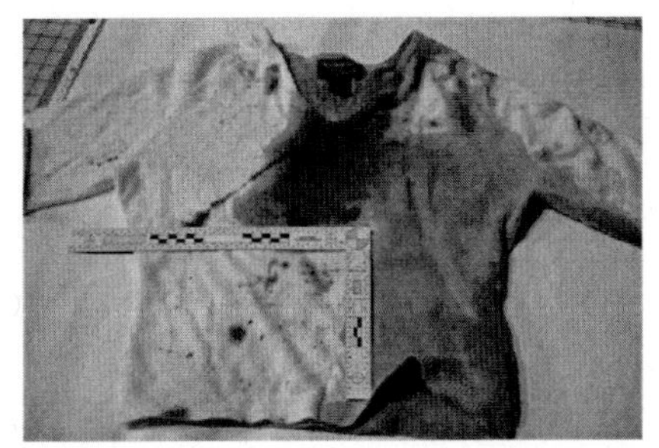

Figures 7.1 and 7.2 Stains on the lower right quadrant of the garment show drip stains; the midline and the lower left quadrant show soaking and wicking stains.

where the slug exited, appeared to be of several different types: they were the typical soaking and wicking stains from exposure to a plentiful blood source, but there were also drip stains and cast-off stains—stains with satellite spatter from vertical drops and stains with tails showing an arc of movement throughout their deposition (see Figures 7.1 and 7.4).

This meant that the various stains observed on the front of the sweater resulted from different mechanisms. These mechanisms included soaking and wicking, dripping from a blood source higher than the sweater's surface, and droplets cast off from some blood source moving in specific directions above the sweater's surface. So multiple events had occurred to deposit bloodstains on the front of the sweater—no single shooting event alone could explain all these different stains.

Examining the back of the sweater also revealed compelling bloodstain data. Recording overall macro observations is critical, but simply looking at bloodstains with the unaided eye seldom reveals sufficient detail to help classify and distinguish different bloodstain patterns. Unfortunately, the state expert's work did not include microscopic examinations, and consequently included no photomicrographs of any relevant bloodstains. The failure to make adequate or detailed observations becomes a familiar flaw with the so-called cookie-cutter method: There is no point raising any further questions when the answers have already been supplied by the generalization applied.

The back of the sweater presented key pieces of information about the shooting, including the three entrance holes—one from the slug and one from each of the two sides of the sabot. But it also revealed another bloodstain pattern visible only microscopically, a pattern that raised further questions about the ballistics issues involved in the shooting (see Figures 7.5 and 7.6).

The sweater was designed with multiple fuzzy fibers covering the entire white surface as part of its weave pattern—a "shaggy" sweater. Under the microscope, near each of the entrance holes, small droplets of blood appear stuck to these fibers (see Figures 7.7 and 7.8). Recall that when a projectile strikes a blood source, the impact produces small blood drops that travel

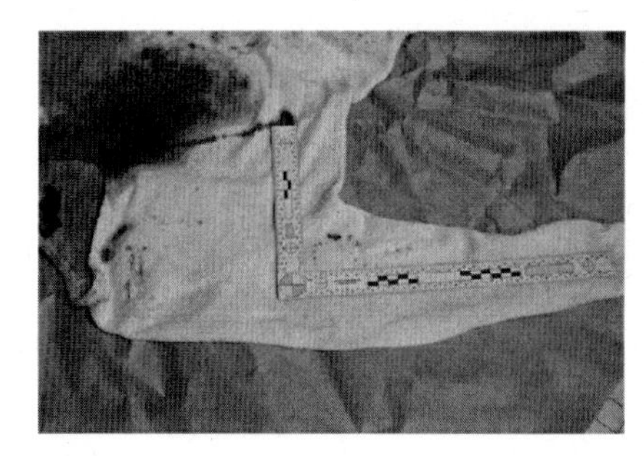

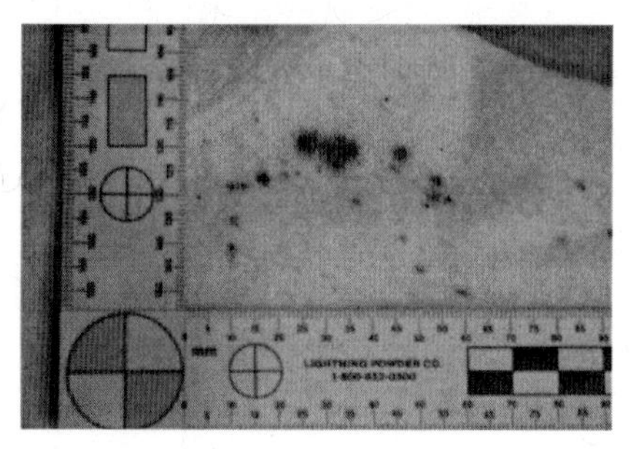

Figures 7.3 and 7.4 Stains on the right arm exhibit directionality as well as a smaller size characteristic of stains cast off from some moving blood source. The photo on the right shows the mixture of these stains—the larger drips with some smaller cast-off stains.

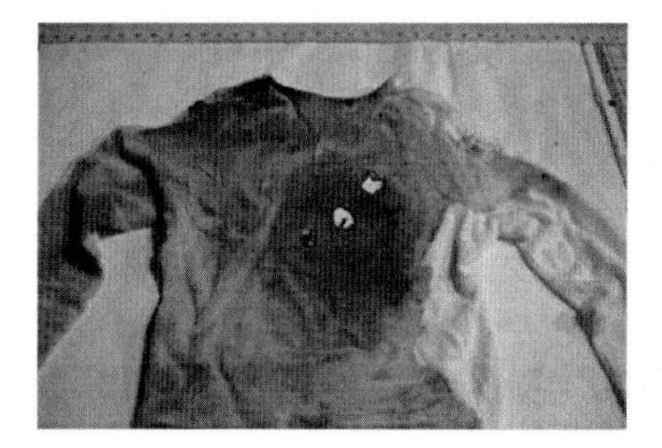

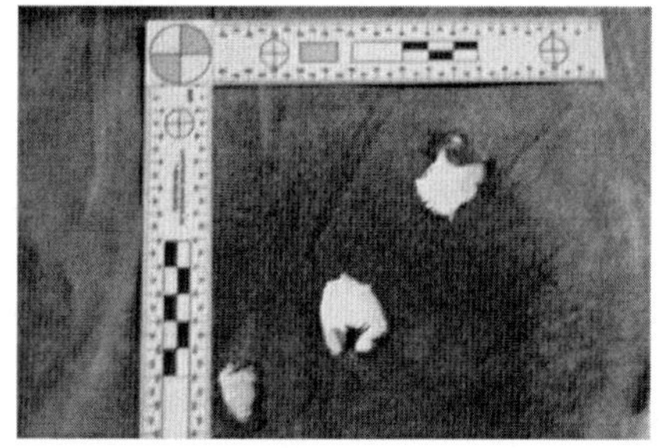

Figures 7.5 and 7.6 The overall views seen here at the left and right show evidence of the soaking and wicking stains as well as the holes in the sweater made by the slug and the sabot, which guided the slug through the 12-gauge shotgun's barrel. Recall that the shotgun itself was never recovered.

back toward the source of the projectile. This phenomenon is called backspatter. It is commonly observed around gunshot wounds of entrance.

If the gun recovered at a shooting scene has small blood droplets present in and around the barrel, these data can be used to help estimate the muzzle-to-target distance. Generally, by experiment, these blood droplets travel backward from the target no more than 20 to 36 inches.* If a shooter has blood droplets on his face, hands, arms, or chest, and if these droplets originate from such backspatter, then we have reason to believe that the shooter's face, hands, arms, or chest was 20 to 36 inches from the blood source when the bullet or projectile struck the target.

* There are two physical phenomena at work here: one is backspatter from impact with the target, as discussed here; the other is a vacuum effect whereby the hot gases exiting from the gun barrel along with the bullet create a suction drawing blood droplets or other materials into the gun barrel itself. Usually small bloodstains inside the gun's barrel originate from this phenomenon rather than from backspatter alone. I use Mikrosil casting material applied inside the gun's barrel to document and recover such bloodstains, if present.

However, these fibers on the decedent's fuzzy sweater served to block this backspattered blood, capturing most of the droplets as they traveled back toward the shotgun's muzzle. Thus, regardless of the actual distance of the shotgun's muzzle or the shooter's face from the victim's back, even at 20 to 36 inches, no backspattered blood would be evident in this case. This demonstrates a common exception to the so-called rule, or generalization, that backspattered blood

Figures 7.7 and 7.8 These photomicrographs show backspattered blood droplets captured by the sweater's fibers. Sometimes these small blood drops can be mistaken for burned and unburned gunpowder. In this case a simple analytical test using x-ray fluorescence spectrophotometry will settle the issue: No nitrates and no other elements characteristic of gunpowder are present in the sample.

will always provide information about muzzle-to-target distances involved in any shooting. The cookie-cutter method doesn't always cut it.

Interestingly, the data developed here help our additional scientific efforts to understand the shooting. We have no reason to use the backspattered blood to determine muzzle-to-target distances. First, we know that the droplets were blocked by the sweater, and second, we do not have the shotgun to examine for the presence of bloodstains. But we do have some reasons to think that the muzzle-to-target distance is greater than 36 inches. We observed no muzzle effluents on the back of the sweater either macroscopically or microscopically: No gunshot residues, including no burned or unburned gun powder, and no shot shell talc are present on the sweater.

The entrance holes from the slug and two halves of the sabot, and the absence of gaseous renting, tearing, and burning of the sweater, suggest an intermediate range between the muzzle and the target rather than a close contact or near-contact gunshot wound of entrance. As the data develop, we begin to see connections among their characters and the inferences that they can license through the careful application of scientific method.

Scientific Inferences from the Suspect's Glasses

The focus of the state's scientific argument remains the bloodstains observed on the suspect's glasses. The claim that only a gunshot entrance wound can produce this high-velocity backspatter has at least been put on notice by the examination of the sweater in this case. Clearly the data from the sweater show that no backspatter from the entrance wounds could account for these stains. In addition, the data from the sweater show that the gunshot wound is at least an intermediate-range injury rather than a close-range or contact gunshot entrance wound. This also precludes the stains on the glasses originating from high-velocity backspatter.

When faced with a question of the possible origin of bloodstains, a scientific approach suggests some sort of testing or experimentation. When hypotheses are developed, they must be tested. In the cookie-cutter approach to science, merely suggesting some hypothesis to explain some data appears to be sufficient to confirm that hypothesis: These cookie cutters seem to be self-confirming. The small stains on the suspect's glasses suggest high-velocity backspatter, and the hypothesis of high-velocity backspatter is confirmed by the small stains appearing on the glasses.

Fortunately, the natural sciences do not work that way. Hypotheses must be tested independently of their derivations, and the results can be used in various ways depending upon what the test or experiment is designed to accomplish. So experimental results can suggest revisions of the original hypotheses as tested, as well as potentially providing confirmations or refutations of these original hypotheses. If our hypothesis that the small stains on the suspect's glasses originate from high-velocity backspatter does not work, then what does? The suspect himself provides another hypothesis suitable for testing. The scientific method intellectually demands that such tests be conducted.

The suspect's account of his actions at the death scene included approaching the front of the body, bending over to touch her chest, finding that his hands became covered with blood, standing before the body for a second or two with his hands suspended in front of his own chest, and then shaking his hands in the air in an effort to remove the blood. He then wiped his hands on his jeans before he tried to use the telephone.

This suggests an alternative mechanism to account for the stains on the glasses. The potential virtue of the hypothesis holding that the stains on his glasses originate from blood shaken off his hands is that this hypothesis can also account for the stains observed on the front of the victim's sweater. According to this hypothesis, the drip stains came from blood falling from his hands as they were held in front of his own chest and over her sweater while the cast-off stains came from shaking his hands. I developed a simple test using similar glasses and my own blood (see Figures 7.9 and 7.10).

I drew 20 ml of my own blood and soaked it into a sponge surrounded by butcher paper on the lab floor. I put on the glasses. I squatted down, putting both hands on the blood-soaked sponge. I stood up with my hands suspended in front of my chest, letting the blood drip from my hands onto the butcher paper. I then simply looked down at the sponge on the floor and shook my hands. The observations successfully replicated the drip and cast-off bloodstains on both the front of the sweater and the suspect's glasses (see Figures 7.11 through 7.18).

I then wiped my palms on my jeans to conclude the experiment. The wipe stains on my Levis also matched the stains that I observed on the suspect's jeans.

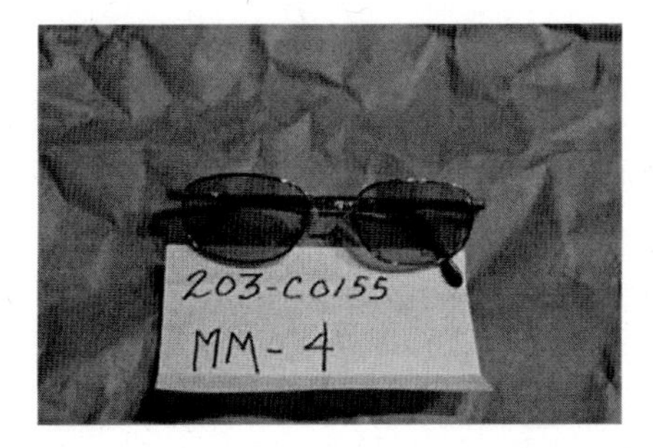

Figures 7.9 and 7.10 The suspect's glasses appear on the left while the test glasses appear on the right. Both were examined with the microscope to analyze the bloodstains present on the lenses.

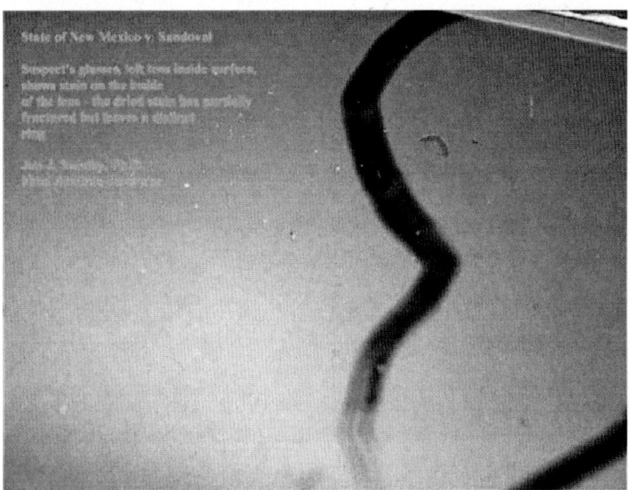

Figures 7.11 and 7.12 The suspect's glasses are seen under magnification. The black line was introduced by the state's examiner—the glasses are dirty and scratched; the stain in the center of the picture is the blood spot. Note how small the stain is—smaller than the diameter of the lead in a modern mechanical pencil.

The more data that the hypothesis consistently accounts for, in general, the more reliable the hypothesis becomes. The foundation for this lies in theories of truth, a province of the philosophical discipline known as epistemology, or theory of knowledge. The relevant point is that if the hypothesis is in fact true, then at the very least, all other facts in the world are logically consistent with that hypothesis.

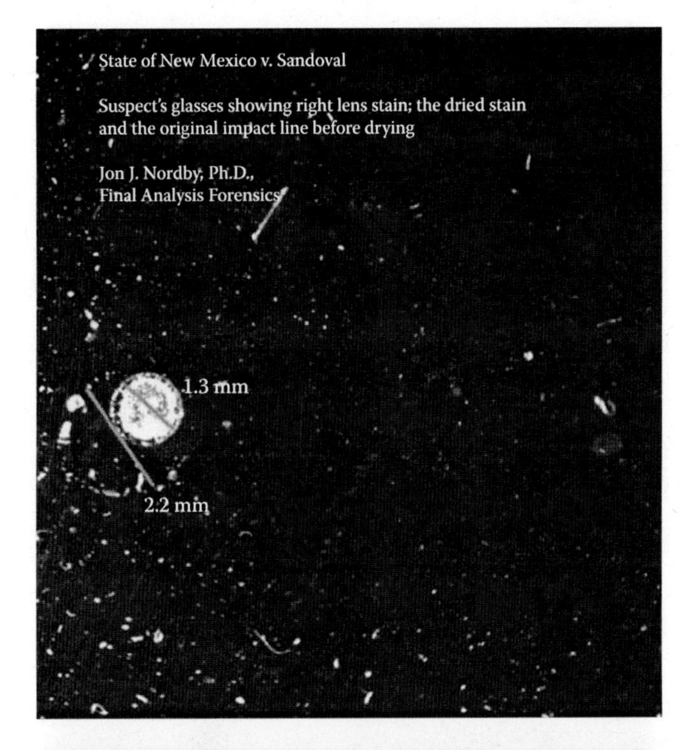

Figures 7.13 and 7.14 This slide shows an interesting characteristic of blood. Notice the small white spot of blood—it is 1.3 mm of blood. However, look closely behind it; you will see the original blood spot, which is 2.2 mm—almost twice as big. As the big spot dried, it shrunk and left the smaller spot behind. The reason this happened is that the serum separates from the red cells and evaporates. The red cells then clump together and clot. The 1.3 mm spot is that clot of red cells. The photo on the bottom shows how small these stains are. (Remember my pencil lead example.) These are the smallest blood spots on the defendant's glasses.

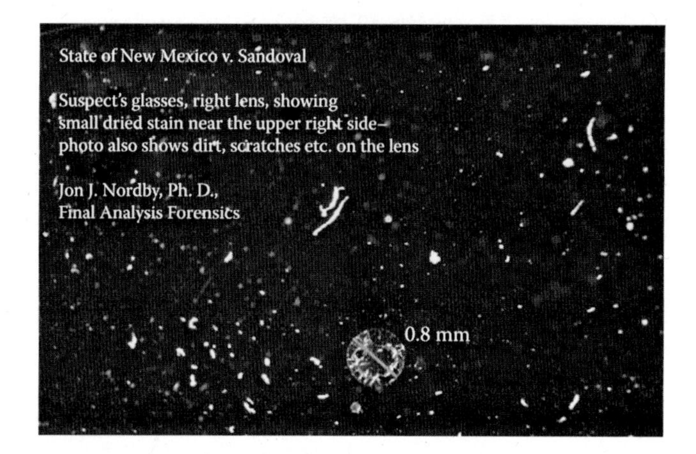

Figure 7.15 This is another example of the small size of the spots present on the glasses.

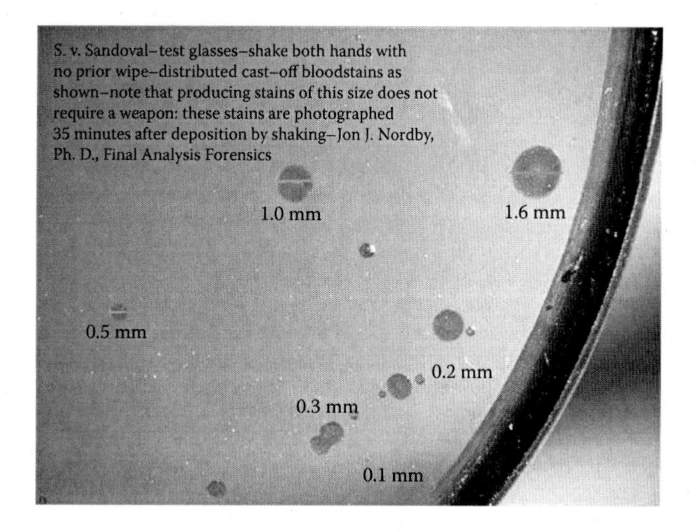

Figure 7.16 Here are blood spots produced by shaking hands that were not first wiped off. Notice the size. Some are bigger and some are smaller. These spots were photographed 35 minutes after the bloody hands were shaken. The spots have not yet started to clot and shrink. It takes approximately 4 hours for them to clot and shrink, depending upon conditions.

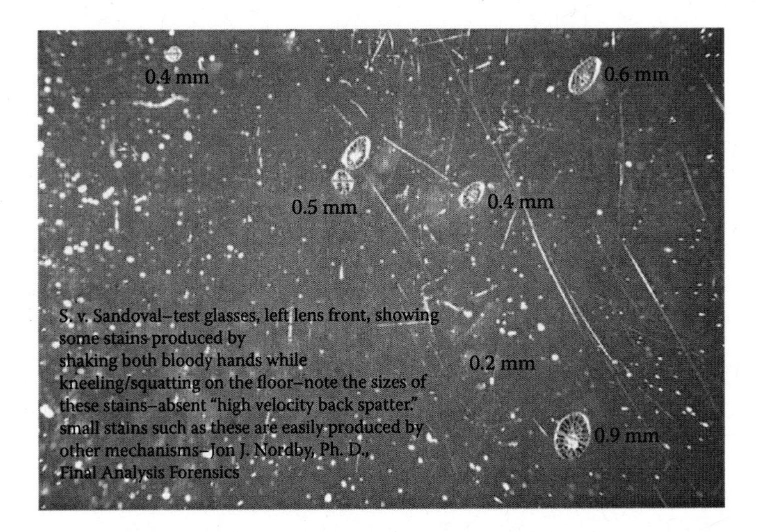

Figure 7.17 To get an accurate result I used test glasses that were also scratched and dirty. The spots were produced by shaking both bloody hands while squatting on the floor. Remember the size of the spots on the defendant's glasses. They ranged from 2.2 to .04 mm in size.

The spots in Figure 7.17 range from .9 to .02 mm in size and they are remarkably similar in size. In fact, there is no relevant difference between these spots and the spots on the defendant's glasses.

As the reader can see, the spots are identical to the spots on the suspect's glasses and my test glasses spots were *not* produced by high-velocity backspatter from a gunshot. They were produced only by shaking bloody hands. There are other ways to produce small spots than high-velocity backspatter, as my experiment has shown. The conclusion, then, is that the spots on the defendant's glasses could have been produced by hand shaking, the way I produced these spots on the test glasses. This is exactly what the suspect reported doing.

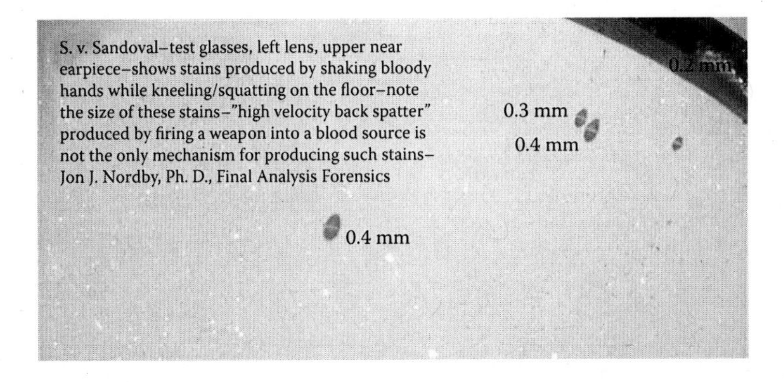

Figure 7.18 Test spots produced by shaking bloody hands.

The assertion that bloodstains as small as 1 mm *must* result from a gunshot is simply proven to be false by this experiment. The steps involved in developing the hypothesis that hand shaking could explain the stains on the glasses involved microscopic examination of the glasses, in turn showing that some of the stains were indeed larger than 1 mm before they dried. The method of proposing, developing, and testing this hypothesis produced results that not only confirmed that such hand shaking *could* account for the stains on the suspect's glasses, but also has the added bonus of accounting for the stains observed on the front of the victim's sweater.

Scientific Conclusions about the Shooting Events

The dependence of good natural science upon data and its cogent development cannot be overemphasized. The state's reconstruction suffers from multiple instances of systematically ignoring significant data. For example, in addition to the problem of flawed bloodstain pattern analysis, if the suspect arrived home when witnesses indicated, and if he did the things that he stated, immediately phoning 911 when arriving at the neighbor's home, and if, as we do know, he stayed on the phone until police arrived, how then did he have time to stage the scene, hide the murder weapon, and dispose of the missing items from the home, let alone find time to shoot the victim? These questions must be raised. Once raised, they require at the very least some effort to provide synoptic answers.

The shotgun was never found despite an extensive search. Nor were the items missing from the home. Such data cannot be systematically ignored just because a cookie-cutter-covering law-like generalization suggests a solution to the case that is independent of these facts. Such pseudoscientific generalizations must never be taken at face value—doing so betrays the very essence of scientific method and flies in the face of its successful practice.

A crime scene reconstruction remains only as reliable as the sciences applied to the data. The sciences applied to the data are only as reliable as the completeness of the recognized, collected, documented, and preserved data. And the data can only be developed through the natural sciences, applied according to the scientific method, which distinguishes reliable scientific inferences from prejudices, lucky guesses, and mere random chance.

Reading List

8

Nihil novi sub sole.
(There is nothing new under the sun.)

Ecclesiastes, Old Testament

Books and Monographs

Forensic Science: An Introduction to Scientific and Investigative Techniques, edited by Stuart James and Jon Nordby, CRC Press, Boca Raton, FL, 2004, 689 pp. Besides being an excellent forensic science survey textbook, and a mystery writer's desk reference, a short but excellent description of the scientific method is on pp. 8 and 9. In this regard, please pause and well consider footnotes 7 and 10 on p. 10.

Guidelines for Failure Investigation, Task Committee on Guidelines for Failure Investigation, ASCE Press, NY, 1989, 221 pp. While the examples in this guide are from civil engineering, the principles and methodology are excellent. It is succinct and full of helpful advice, especially for inexperienced investigators.

Incident Investigation and Reporting, A. Beare and S. Lutterodt, Electric Power Research Institute 1002077, May 2003, about 80 pp. This monograph describes a model procedure for investigating incidents. The model was prepared by surveying the procedures used at fourteen utilities that were believed to have effective root cause programs. Other sources were also included in the model. A good primer on event cause graphing is provided in Appendix C.

The Fundamentals of Forensic Investigation Procedure Guidebook, J. Reynolds, H. Manger, and R. Samm, Electric Power Research Institute 1001890, March 2003, about 55 pp. While intended for use in investigating switchyard and power utility distribution failures, the principles are easily transferred to other applications. It is intended to be a training manual to introduce methodical forensic investigation methods to utility personnel and covers fault tree analysis, witness interview, data gathering, and reports, and also provides a case study. Several useful checklists are included.

National Transportation Safety Board Aviation Investigation Manual—Major Teams Investigations, November 2002, 53 pp. This is the procedure and administrative manual used by the NTSB that explains how investigative teams investigate major incidents. Such a team might have a hundred members and several governmental agencies.

NPFA 921: Fire and Explosion Investigations, National Fire Protection Association, 1992 ed., 120 pp. As its name indicates, this is a brief but useful handbook on how to conduct the investigation of a fire or explosion. Chapter 2 has a good discussion of using the scientific method to determine the cause of a fire or explosion.

NPFA 907M: Determination of Electrical Fire Causes, National Fire Protection Association, 1988 ed., 40 pp. This is a companion to NPFA 921 and is focused on fires caused by electrical faults and failures. Part I is a good overview on how to conduct an investigation, and Chapter 8 in Part II, while specific for electrical fires, provides a good outline of what is expected of an investigator.

Event and Causal Factors Analysis, J. Buys and J. Clark, Technical Research and Analysis Center, SCIENTECH, Inc., SCIE-DOE-01-TRAC-14-95, August 1995, 13 pp. This monograph contains a brief description of the event and causal factors analysis charting method.

Fault Tree Handbook, W. E. Vesely, F. F. Goldberg, N. H. Roberts, and D. F. Haasl, U.S. Nuclear Regulatory Commission, NUREG-0492, January 1981, 209 pp. This publication is essentially an entire course in fault tree analysis theory and application. Chapter VII has a nice explanation of Boolean algebra and how it can be applied to fault trees. This publication can be downloaded from the Internet free of charge in PDF format.

Machinery Failure Analysis and Troubleshooting, 3rd ed., Vol. 2 of the *Practical Machinery Management for Process Plants* series, Heinz Bloch and Fred Geitner, Gulf Publishing, Houston, TX, 1997, 667 pp. Chapters 10 through 13 and Appendixes A and B are especially germane to failure analysis work as discussed in this book, although the entire book is certainly worth the time to read, as it is written to be practical and immediately usable. The seven-cause category approach introduced in Chapter 10 is a good method to apply when a person is otherwise inexperienced in conducting root cause investigations and the failure involves equipment or systems of equipment.

The Human Performance Evaluation Process: A Resource for Reviewing the Identification and Resolution of Human Performance Problems, Valerie Barnes and Brian Haagensen, Office of Nuclear Regulatory Research, NUREG/CR-6751, September 2001, about 125 pp. As the title states, it is a good text that describes how human performance issues are sorted out. Be sure to check out Chapter 7, especially

Figure 7.1, the HPEP cause tree. It is a useful method for quickly sorting out the type of human performance problem that may be involved in a failure.

Crime Scene Investigation: A Guide for Law Enforcement, Research Report, U.S. Department of Justice, National Institute of Justice, NCJ 178280, January 2000, 47 pp. This is a good overview of the recommended practices for preserving crime scene evidence and conducting preliminary investigative work at the crime scene.

Fire and Arson Scene Evidence: A Guide for Public Safety Personnel, Research Report, U.S. Department of Justice, National Institute of Justice, NCJ 181584, June 2000, 66 pp. This publication, much of which was developed at the University of Central Florida, is an attempt at establishing a protocol for gathering evidence at the scene of a fire.

A Guide for Explosion and Bombing Scene Investigation, Research Report, U.S. Department of Justice, National Institute of Justice, NCJ181869, June 2000, 51 pp. As with its sister publication, the University of Central Florida also contributed substantially to this one. This is also an attempt at establishing a protocol for gathering evidence at the scene of an explosion or bombing.

Root Cause Analysis, Good Practice, Institute of Nuclear Power Operations, OE-907, INPO 90-004, January 1990, 74 pp. This root cause how-to manual was developed from surveying a number of nuclear power utilities. The document consists of key principles and examples of methods. The term *root cause* is defined in this document as "the fundamental cause(s) that, if corrected, will prevent recurrence of an event or adverse condition."

Assessment of Accident Investigation Methods, Steve Munson, MS thesis, University of Montana, 1999, 49 pp. The thesis is based upon a case study of wild land fire fighting and how three investigative methods were applied. Two methods were found to be successful: sequential timing and events process (STEP) and fault tree analysis. The third method, controls/barriers analysis, was found to be not applicable to the case study that involved wild land firefighter entrapments. Within the thesis is an excellent comparison of the various investigative methods, including MORT, noting their strong points and weak points. Mr. Munson did a fine job.

Procedures for Performing a Failure Mode, Effects and Criticality Analysis (FMECA), MIL-STD-1629A, November 24, 1980. This is the formal procedure used by all departments and agencies of the U.S. Department of Defense. As with many military publications, this one can be hard to follow at first if the reader has not previously been exposed to military procurement jargon. The document lays out the military method of determining potential failure modes in a system and the resulting consequences. If used in reverse, the method can be useful in diagnosing failures in systems.

Guidelines for the Naval Aviation Reliability-Centered Maintenance Process, NAVAIR 00-25-403, March 1, 2003. RCM is the acronym for reliability-centered maintenance. This is an analytic process to determine preventive maintenance requirements. The methodology is used to prevent failures from occurring by considering how failures might occur and what to do to prevent them from occurring. The heart of this method is FMECA, failure modes, effects, and criticality analysis. This differs from a basic FMEA, failure modes and effects analysis, in that it also considers the consequences of the failure.

On Man in the Universe, Introduction by Louside Loomis, Walter Black, Inc., Roslyn, NY, 1943. There is a brief but good discussion of Aristotle's method of inductive logic on p. xiv.

Reason and Responsibility, edited by Joel Feinburg, Dickenson Publishing, Encino, CA, 1971. The *a priori* method of reasoning is discussed at length in the essay authored by A. J. Ayer, "The A Priori," beginning on p. 181.

The Columbia History of the World, edited by J. A. Garraty and P. Gay, Harper and Row, NY, 1981. Pages 681–686 discuss Aristotle's inductive method.

General Chemistry, Linus Pauling, Dover Publications, NY, 1970. Pages 13–15 discuss the modern scientific method as espoused by Dr. Pauling, a Nobel Prize laureate.

The Engineering Handbook, edited by Richard Dorf, CRC Press, Boca Raton, FL, 1995. Pages 1897–1906 discuss briefly several systematic approaches to failure and hazard analysis.

To Engineer Is Human, Henry Petroski, Vintage Books, 1992, 251 pp. This book is about failure with respect to engineering design and has an excellent chapter about forensic engineering (Chapter 14). I highly recommend this book for its insights into the workings of engineering failure. It is also very readable.

Bones, Douglas H. Ubelaker and Henry Scammell, Harper Collins, NY, 1992, 317 pp. This is a casebook of a forensic anthropologist. The book is not only informative, but also interesting to read. Dr. Ubelaker reconstructs who a person was from his remains by applying what has been already learned in physical anthropology and medical science and then working backwards.

The Seashell on the Mountaintop, Alan Cutler, Plume Books (Penguin Group), NY, 2003, 228 pp. Interesting by itself as a biography of one of the post-Galilean Renaissance scientists in the seventeenth century, Nicolaus Steno, it is an absorbing story of a person tenaciously applying the scientific method to explain one problem, why seashells are found on mountaintops, and by doing so inventing a new science, geology. This is an excellent book that exemplifies the conflicts of a worldview based upon deductive reasoning and religious dogma versus a worldview based on inductive reasoning and accepting evidence at face value.

Managing the Risks of Organizational Accidents, James Reason, Ashgate Publishing, Burlington, VT, 1997, 252 pp. This book assesses how failures occur in organizations, especially how to detect organizational and process weaknesses. I recommend it to managers of power plants, manufacturing facilities, hotels, hospitals, ships, and other organizations that have the potential to cause harm to the public. It has some excellent recommendations about how to avoid organizational failure.

Galileo's Revenge, Peter Huber, Basic Books, NY, 1991. This book discusses the apparent proliferation of "junk science" testimony in the court room in many high-profile cases.

The New Rational Manager, Charles Kepner and Benjamin Tregoe, reissued by Princeton Research Press, Princeton, NJ, 1981, 224 pp. This is the text upon which the K-T method is based.

Forensic Engineering, edited by Kenneth Carper, Elsevier, NY, 1989, 353 pp. This is an introductory text that explains and defines forensic engineering. There are fourteen chapters, each written by a different author, which describe some facet of forensic engineering. Chapter 3, authored by Paul E. Pritzer, PE, is about forensic engineering with respect to fires and discusses briefly the investigative team concept. I also recommend Chapter 10 (photo documentation), Chapter 11 (reports), and Chapter 12 (expert witness).

Root Cause Analysis Guidance Document, U.S. Department of Energy, DOE-NE-STD-1004-92, February 1992, about 30 pp. As the name implies, this document outlines the basic steps used by the Department of Energy in conducting a root cause assessment. In Table 1 is a handy comparison of the strengths and weaknesses of various investigative methods.

Root Cause Analysis: Improving Performance for Bottom-Line Results, Robert Latino and Kenneth Latino, 3rd ed., CRC Taylor and Francis, Boca Raton, FL, 2006, 266 pp. This book about root cause is aimed at persons in management. It discusses at length how to initiate and maintain a root cause program in a large organization to reduce internal costs. However, the lessons concerning the management of a root cause organization can also be applied to most organizations that regularly perform scientific investigations of failure events or crimes.

Introduction to Mathematical Statistics, Paul Hoel, John Wiley & Sons, NY, 1971. Pages 190–224 discuss general principles for statistical inference, one of the methodologies employed in modern inductive reasoning.

Reporting Technical Information, Houp and Pearsall, Glencoe Press, Beverly Hills, CA, 1968. With respect to the topics discussed, pp. 224–225 provide a format used for the reporting of physical research findings. Pages 47–56 provide an excellent discussion of audience analysis, and pp. 57–68 provide useful information for the organization of findings and observations. Overall, this text provides an excellent foundation for technical writing.

The Elements of Style, by Strunk and White, 4th ed., Longman Publishers, NY, 2000, 105 pp. As one reviewer said of this book, "It's hard to imagine an engineer or a manager who doesn't need to express himself in English prose as part of his job." Chapter V, "An Approach to Style," beginning on p. 66, provides a short list of writing style reminders that certainly apply to reporting the findings of a failure investigation.

Murders in the Rue Morgue, Edgar Allen Poe, 1841. There are several editions extant. Several are even available free of charge on the Internet. This is an interesting novella, as it is considered the first modern detective story. What is interesting with respect to failure analysis is that in the first two pages there is an excellent exposition of how to conduct an impartial, fact-based failure investigation. The novella then gives an example of a case that would challenge normal investigation methods, and then concludes by explaining how the method was used to solve this otherwise inexplicable case.

Fingerprints, Colin Beavan, Hyperion, NY, 2001, 232 pp. Excellent nonfiction book that provides a history of fingerprints and information about fingerprints and fingerprinting, and tells the interesting case story that introduced fingerprinting as a *bona fide* forensic technique.

Inviting Disaster: Lessons from the Edge of Technology, James R. Chiles, Harper Business, NY, 2002, 338 pp. Well-written narrative stories about how some relatively recent disasters happened. Some of the events include Chernobyl, Bhopal, the space shuttles, the World Trade Center, the crash of the Concorde, and the *Thresher*. Having them succinctly packaged in one book is very useful, especially for classroom discussion of common factors.

Vidocq: A Biography, John Philip Stead, Ray Publishers, NY, 1954. Vidocq wrote his memoirs in four volumes, which are entitled *Memoires de Vidocq, Chef de Police de Surete Jusqu'en 1827*. However, since it is well recognized that these memoirs were likely all or largely ghost written and generously exaggerated, and that Vidocq only authorized the first two volumes, I recommend Mr. Stead's book instead for a more objective perspective about Vidocq's life. A 1946 United Artists English language motion picture about Vidocq was made starring George Sanders and Carole Landis, entitled *A Scandal in Paris.* It is also distributed under the title *Thieves Holiday.* The movie is more of a comedy and a slightly naughty (for the 1940s) bedroom farce than a genuine depiction of Vidocq. A more recent French language motion picture in 2001 starred Gerard Depardieu as the intrepid detective. It was not well received by the French critics.

The Dream of Reason, Anthony Gottlieb, W. W. Norton, NY, January 2001, 352 pp. The book is a history of philosophy from ancient Greece to the Renaissance. A large portion of the book, however, is devoted to logic, especially the deductive and inductive methods. The chapters that involve the development of the scientific method, William of Occam, and Francis Bacon are especially germane.

Roots of Scientific Thought, Philip P. Wiener and Aaron Noland, Basic Books, NY, 1957, 677 pp. This is an excellent survey of what has constituted scientific thinking and evidence at various times in the history of Western civilization. It is especially notable in showing how social developments have interacted with scientific thinking and vice versa. While the articles are written in an academic style that can sometimes be difficult, the content is excellent.

How We Know What Isn't So, Thomas Gilovich, Free Press, NY, 1991, 216 pp. This is a readable discussion of why people believe irrational things and the nature of bias in the interpretation of data.

Failure Mode and Effect Analysis, 2nd ed., D. H. Stamatis, ASQ Quality Press, Milwaukee, WI, 2003, 455 pp. plus CD supplement. This college text explains well the how-to mechanics of creating FMEA models for a wide variety of industries.

The Basics of FMEA, McDermott, Mikulak, and Beauregard, Productivity Press, Florence, KY, 1996, 75 pp. While this small handbook does not offer the breadth of industry examples as the text by Stamatis, it does provide an easy-to-read explanation of how FMEA works and what to do to create one.

Columbia Accident Investigation Board, Report Vol. 1, August 2003, available at http://caib.nasa.gov/news/report/volume1/default.html. This is a sizable download for all the volumes. A readable overview is contained in Vol. 1, which will download in a PDF format at about 11 MB, or also in smaller sections. It is also available in a paper version with a CD-ROM through the U.S. Government Printing Office, NASA, and the National Technical Information Service. This is possibly the most thorough and detailed analysis of a major failure event readily available to the public. It is a classic study of how even a multilayered barrier scheme put together by the best and brightest can be circumvented by a latently flawed decision-making process. This should be required reading by all senior managers in industries that hold the public trust, such as the nuclear industry, the defense industry, and the chemical industry. Unfortunately, it is not.

Papers and Articles

"Trial and Error," Harry Saunders and Joshua Genser, *The Sciences*, September/October 1999, Vol. 39, No. 5, pp. 18–23, published by the New York Academy of Sciences. This is an interesting short article about defining *reasonable doubt* in terms of decision theory and uncertainty. The authors discuss how the vagueness of the legal term *reasonable doubt* can lead juries to different verdicts despite the fact that they hold similar beliefs concerning the defendant's guilt.

"When Is Seeing Believing?" William Mitchell, *Scientific American*, February 1994, Vol. 270, No. 2, pp. 68–75. This is recommended reading for all persons involved in the presentation and examination of evidence. The article describes some of the ways in which computer graphics can be used to provide fake photographic evidence. Unfortunately, one of the darker sides of the computer revolution is that photographic evidence, once considered quite reliable, is now becoming increasingly easier to fake. It will not be long until unscrupulous forensic experts use computerized faking techniques to persuade juries as to the "scientific" evidence in a case, especially when the monetary stakes are sufficiently high.

"Chemist in the Courtroom," Robert Athey, Jr., *American Scientist*, September-October 1999, Vol. 87, No. 5, pp. 390–391, published by Sigma Xi. This is a short article describing a forensic chemist's experiences as an expert witness.

"Register or Perish," Marina Krakowsky, *Scientific American*, December 2004, pp. 18–19. The article discusses the publication bias in medical journals, wherein research that reports positive information about a drug is published, but research that finds negative information or neutral information about a drug is not published.

"Improve Your Root Cause Analysis," Duke Okes, CMC, *Manufacturing Engineering*, March 2005, pp. 171–178. The article outlines the general method of root cause analysis used in the manufacturing industry and provides some useful tips. An excellent point made early in the paper is "root cause analysis is not the same as problem solving. Root cause analysis is only the diagnostic part of the problem solving process, but it's the part that if done incorrectly, causes all subsequent actions to have little value."

"Punishing People or Learning from Failure? The Choice Is Ours," Sidney Dekker, Centre for Human Factors in Aviation, IKP, Linkoping Institute of Technology, Sweden, at http://www.hufag.nl. Excellent paper describing the tendency to punish people who make errors that lead to failures rather than understanding the context of why the error was made and correcting it.

Title 10, Courts and Judicial Procedure, Part III, Procedure, Chapter 43, Evidence and Witnesses, Subchapter III, Chain of Custody: http://delcode.delaware.gov/title10/c043/sc03/index.shtml. These are the legal requirements in the state of Delaware with respect to chain of custody.

"Believe It or Not: The Battle over Certainty, A Point of View," Lisa Jardine, BBC News, April 28, 2006, published on the BBC News website, http://news.bbc.co.uk/2/hi/uk_news/magazine/4950876.stm. This is the source of the story used in Chapter 2 about the pendulum clock and the doubts of Mr. Christiaan Huygens, the inventor of the clocks, about their reported accuracy.

"How to Investigate a Process Interruption," Raymond Atkins, CPMM, *Maintenance Technology Magazine*, January 2008, pp. 44–47. There is also a magazine website, http://www.mt-online.com/, on which the article has been archived. This is a succinct explanation of how to investigate problems associated with a process interruption. There is a short section noting the similarity of conducting a root cause failure analysis (RCFA) and a failure modes and effects analysis (FMEA) on p. 47. The succinct advice on p. 46 concerning reviewing PM (preventative maintenance) and PdM (predictive maintenance) work orders and also comparing the performance of the offending machine with another, a variant of change analysis, is excellent. These items are often overlooked.

"Forget about Causes, Focus on Solutions," Fred Nichols, 2004, available at http://home.att.net/~OPSINC/forget_about_causes.pdf. This is a short paper that discusses the history of the gap theory, which is one of the fundamental principles applied in the K-T method. Mr. Nichols makes the point in this paper that often there is little that can be done concerning the causes of problems and that sometimes the only practical thing to do is focus on solutions.

Index

F

R

S

T

U